Discard
2009

INVESTIGATING THE OZONE HOLE

INVESTIGATING THE ZONE HOLE

by Rebecca L. Johnson

Lerner Publications Company ▪ Minneapolis

Acknowledgments

Research for this book was funded by a grant from the National Science Foundation's Antarctic Artists and Writers Program. I wish to thank the members of NSF's Division of Polar Programs who supported and encouraged my work, as well as the many scientists, staff, and military personnel who assisted me during this research.

Library of Congress Cataloging-in-Publication Data

Johnson, Rebecca L.
    Investigating the ozone hole / by Rebecca L. Johnson.
        p. cm.
    Includes index.
    Summary: Discusses the scientific studies of the ozone layer in
the Earth's atmosphere, and the causes and effects of depletions of
this layer.
    ISBN 0-8225-1574-1
    1. Chlorofluorocarbons—Environmental aspects—Juvenile
literature. 2. Ozone layer depletion—Juvenile literature.
3. Ultraviolet radiation—Environmental aspects—Juvenile
literature. [1. Ozone layer. 2. Ozone layer depletion.]
I. Title.
TD887.C47J65   1993
551.5'112—dc20   93-15225
                                                              CIP
                                                               AC

Manufactured in the United States of America
2 3 4 5 6 - P/JR - 99 98 97 96 95

CONTENTS

Fierce ground blizzards and subzero temperatures are part of the challenge of doing scientific research in Antarctica.

INTRODUCTION

Leaning back in his chair, Joe Smith stopped working for a minute and listened to the wind outside. It was really howling today, a nonstop roar that made the tiny building shudder. Joe got up and looked out the window. On the other side of the frosty glass, he could see snow blowing past. Beyond that, there was nothing but gloomy darkness. The thermometer that kept track of the temperature outside read −40°C (−40°F). He looked at his watch: 11:45 A.M. It was nearly time for lunch, but a breath of fresh air would feel good first.

Joe slipped on a down vest over his wool sweater and then pulled on a pair of thick insulated pants over his jeans. He stepped into a pair of huge white boots and tied the ankles of the pants tightly over the boot tops. Next he struggled into a big down parka and coiled a Polarplus muffler snugly around his neck. Finally he pulled a wool face mask over his head, put on a pair of ski goggles, and stuffed his hands into two pairs of heavy mittens. All zipped up and buttoned down, he opened a heavy door at one end of the room and stepped into a cold, dark hallway.

Frost decorated the hallway walls. As Joe exhaled, his warm, moist breath billowed out as a cloud of vapor in front of his face. He stomped down the hallway, pulled open the door at its end, and stepped out into Antarctica's frigid winter darkness.

The wind nearly pushed him over. It was so cold that it was hard to breathe. Blowing snow made it impossible to see ahead or behind. But overhead, Joe could see stars and the full moon. Then, as he watched, colored bands of light suddenly rippled and flashed across the sky. Seeing the aurora australis, or southern lights,

The aurora australis frequently lights up Antarctica's dark winter sky.

Joe forgot about the cold for a minute or two. With the wind raging around him, he stood quietly and watched a spectacle few people have ever seen firsthand.

Joe Smith is a research technician working in Antarctica. It is now July, the middle of the winter on this frozen continent at the bottom of the world. Joe arrived in February. He flew in on a large military cargo plane and landed near McMurdo Station, the United States' largest scientific research station in Antarctica. McMurdo is located on Ross Island, a barren volcanic island that lies just off the Antarctic continent at a latitude of 78°S. The station consists of several dozen buildings, piles of cargo, huge storage tanks of fuel, and lots of heavy equipment. McMurdo is an important center for the U.S. Antarctic Program. During the summer season—October to February—there are usually more than 1,200 scientists and support personnel working there. Only about 200 hardy souls, however, stay in McMurdo during the winter months.

Joe is one of those hardy individuals. He is the "winter-over" member of a science research group from the U.S. National Oceanic and Atmospheric Administration (NOAA). People who spend the winter in Antarctica not only have to deal with fierce

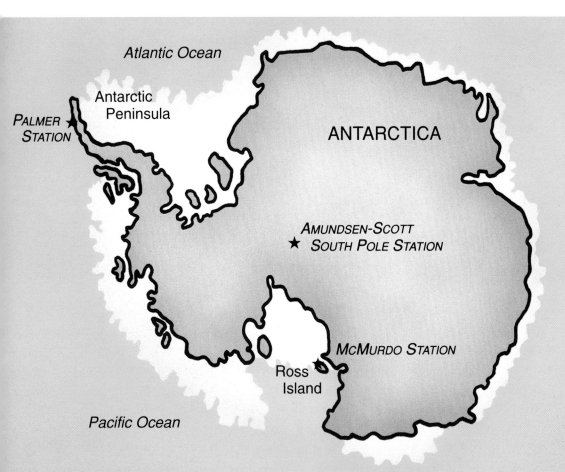

A map of Antarctica showing the three major scientific research stations operated by the U.S. Antarctic Program: McMurdo Station, Palmer Station, and Amundsen-Scott South Pole Station

A view of McMurdo Station, the United States' largest scientific research base in Antarctica

cold, as Antarctica is the coldest place on earth, they must also put up with nearly six months of darkness. From March until early September, the sun never rises above the horizon. It takes a special kind of person to live and work in such a harsh environment.

The other members of the NOAA science team will arrive in McMurdo in late August. Until they do, Joe will carry out the work alone. Most of his time is spent where he is now, at a tiny field research hut about 4 kilometers (2.5 miles) from McMurdo. The hut, surrounded by a tangle of wires and aerials, is perched on a bare, windswept volcanic ridge called Arrival Heights. Arrival

Inside the hut at Arrival Heights, Joe Smith monitors data coming from a Dobson spectrophotometer.

The research hut at Arrival Heights can be a lonely scientific outpost.

Heights isn't the kind of place where most people would want to spend much time. But it is the site of some very important scientific research.

Even through many layers of clothing, Joe soon felt cold and headed back inside the hut. After a quick lunch, he went back to work. The small room was filled with equipment and computers. He sat down in front of one computer terminal and looked at the numbers appearing in several columns on the monitor screen. The computer was receiving data from a Dobson spectrophotometer, an instrument that is used to measure amounts of various substances, primarily gases, in the atmosphere.

To take measurements with a Dobson, you need sunlight. You can get that sunlight in two ways: directly, from the sun when it is up during the day, or indirectly, from the moon. The moon doesn't produce light of its own. What we call moonlight is simply sunlight that is reflecting off the surface of the moon.

From his cold and lonely post at Arrival Heights, Joe Smith has spent much of the dark Antarctic winter using moonlight to take measurements with the Dobson. Down in McMurdo, Joe quickly earned the nickname "Moon Man" Smith because of the work he was doing. Whenever the moon was visible and at least half full, Joe was at Arrival Heights using the Dobson, with the help of computers, to measure **ozone** in the atmosphere high overhead.

Ozone is a type of gas that forms a layer—the **ozone layer**—high above the earth. The ozone layer has been a part of the earth's atmosphere for millions of years. But in the last 10 years or so, scientists have discovered that the ozone layer is in trouble. Early each spring in Antarctica, when the sun comes up over the horizon, ozone begins to disappear above the continent. Within a few weeks after the sun first rises, so much ozone disappears that a "hole" forms in the ozone layer over Antarctica.

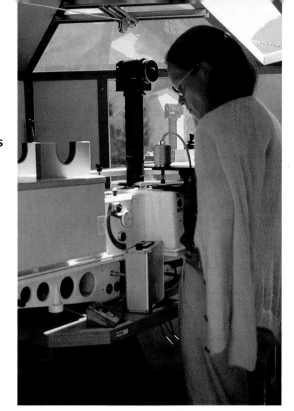

A Dobson spectrophotometer uses light to measure ozone and other substances in the atmosphere. Light enters the instrument through the periscope-like structure at the top.

You may have heard about ozone destruction and Antarctica's ozone hole. Ozone destruction has been in the news a lot lately. But it's not only above Antarctica that ozone is disappearing. The entire ozone layer, all around the planet, is at risk. Damage to the ozone layer is a serious environmental problem facing the world today.

Joe Smith and ozone researchers from many different countries are working hard, often under extremely harsh conditions in remote places, to understand this problem of ozone destruction. In the following pages, we will take a look at what scientists have discovered about the Antarctic ozone hole and global ozone loss—or ozone depletion, as it is often called. But to do so, we first need to answer a few basic questions. For instance, what is ozone? Where is it found? And why is it so important that researchers travel to places like Antarctica to study it?

Like a blanket of air, earth's atmosphere surrounds the planet and makes life on the surface possible.

14

CHAPTER

1

Ozone and the Ozone Layer

The earth's atmosphere is a blanket of air that surrounds the planet. Air is a mixture of many different gases. The two gases that are the most plentiful are nitrogen and oxygen. About 78 percent of the volume of the atmosphere is nitrogen, while slightly less than 21 percent is oxygen. The remaining 1 percent or so is made up of at least a dozen other gases, including carbon dioxide, helium, and ozone, to name just a few. In addition, the atmosphere contains water in three different forms: as tiny crystals of ice, as liquid droplets, and as invisible, gaseous water vapor.

The atmosphere extends many kilometers out from the earth's surface. But it is not a uniform layer of gas from top to bottom. Because of the pulling force of gravity on the molecules of gas that make up the air, the atmosphere is much more dense—the gas molecules are more numerous and closer together—near the ground than it is high overhead. Moving out from the planet's surface, the atmosphere becomes progressively less dense. This means there are fewer gas molecules, and they are farther apart, the greater the distance from the earth.

The earth's atmosphere can be divided into four major regions. But the boundaries between the regions are not very distinct, so

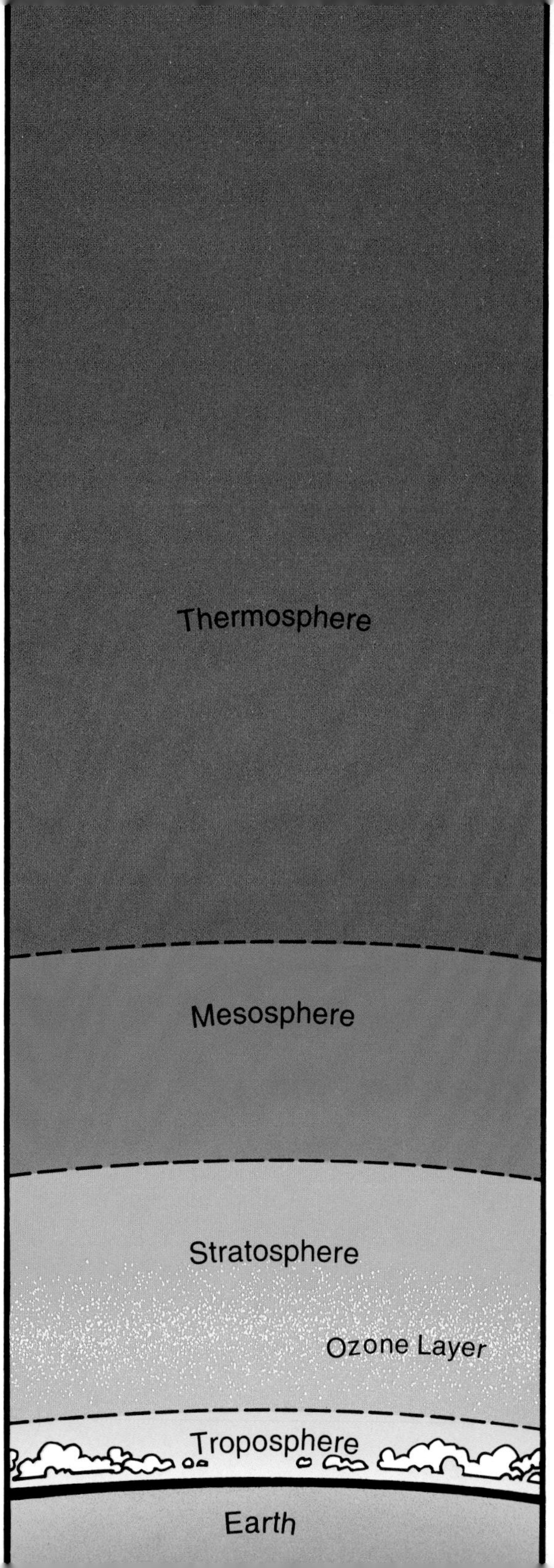

The regions of the earth's atmosphere

16

Thermosphere

Mesosphere

Stratosphere

Ozone Layer

Troposphere

Earth

we can say only approximately where one region ends and the next one begins.

The first region of the atmosphere, closest to the earth, is called the **troposphere.** It extends about 10 kilometers (roughly 6.5 miles) above the planet's surface. The troposphere contains the air we breathe and a considerable amount of water vapor. It is in this region where most of the earth's weather—clouds, thunderstorms, hurricanes—takes place.

Beyond the troposphere is the second region of the atmosphere, the **stratosphere.** The boundaries of the stratosphere extend from roughly 10 to 48 kilometers (6.5 to 30 miles) up. The air in the stratosphere is much less dense than it is in the troposphere, and it is much drier (there is very little water vapor). High-flying jets cruise through the lower part of the stratosphere, and some types of rare, high-altitude clouds are found there. But the stratosphere is best known for containing the ozone layer.

Beyond the stratosphere are the **mesosphere,** which extends to 80 kilometers (50 miles) above the planet, and finally, the **thermosphere.** The outermost edge of the thermosphere is roughly 965 kilometers (600 miles) above the earth's surface. Beyond it, the airless vacuum of space begins.

With each breath, you inhale molecules of oxygen gas. Oxygen is something that all the cells in your body need constantly to stay healthy and alive. The same is true for all other animals, including fish and other aquatic animals. Oxygen from the air dissolves in water and enters the bodies of aquatic animals through gills or similar structures. Plants, too, must have oxygen in the immediate environment to survive.

If we could somehow magnify one molecule of oxygen gas, we would see that it is made up of two oxygen atoms that are bonded together. Scientists use a capital letter O to represent a single atom of oxygen. Therefore, the symbol for a molecule of oxygen gas—two oxygen atoms bound together—is O_2.

Like oxygen, ozone is a gas that is made up of oxygen atoms. But a molecule of ozone contains three atoms of oxygen bonded together (O_3). Just one more oxygen atom might not seem to be enough to make much of a difference, but it is. Oxygen and ozone are extremely different substances. For instance, breathing ozone can be harmful. Unlike oxygen, ozone is poisonous to cells. People and other animals become sick and may even die if they breathe ozone gas for long. And yet, ozone in the atmosphere is absolutely essential for the well-being of almost all living things on the earth.

Compared to nitrogen and oxygen, the amount of ozone in the atmosphere is very small—it makes up only 0.01 percent of the atmosphere. About 90 percent of all ozone is found in the stratosphere, where it is concentrated in a layer between 12 and 35 kilometers (roughly 7 and 22 miles) above the ground.

The thickness of the ozone layer makes it seem like there is a lot of ozone up there. But there's actually not all that much— in the stratosphere, gas molecules are spread out over a lot of space. If you could somehow bring the ozone layer down to ground level, so that its molecules were as close together as air molecules are near the earth, the ozone layer would be just a few millimeters thick!

So why is most of the atmosphere's ozone concentrated in the stratosphere? The answer is that ozone is made there. In the stratosphere, ozone is continually being formed, broken down, and then reformed, over and over again. Three key ingredients are involved in this stratospheric ozone cycle: oxygen, ozone, and energy from the sun.

The sun is the ultimate source of energy for our planet. Solar energy is produced by explosive nuclear reactions that take place continuously in the interior of this fiery star. Huge amounts of solar energy constantly radiate from the sun's surface. This energy travels through space in the form of **electromagnetic radiation.**

One way to understand electromagnetic radiation is to think of it as "energy waves." Scientists often describe waves of electromagnetic radiation in terms of their different wavelengths. What does it mean when we talk about a wave's length? If you toss a pebble into a pond, small waves are created that spread out from the point where the pebble hit the water's surface. The up-and-down motion of the water in these waves forms a series of high points and low points. The distance from one high point to the next high point (or from one low point to the next low point) is the **wavelength** of that wave. Like water waves, waves of energy also have high points and low points, and their wavelengths are measured in the same way.

The waves of electromagnetic radiation that are emitted by the sun have a wide range of wavelengths. This range is known as the **electromagnetic spectrum.** At one end of the spectrum are gamma rays, which have extremely short wavelengths. At the other end of the spectrum are radio waves, which have very long

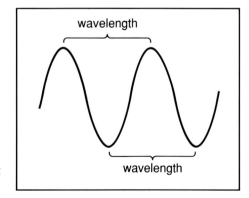

The length of a wave is measured from one high point to the next high point, or from one low point to the next low point.

The Electromagnetic Spectrum

Gamma rays have the shortest wavelengths on the spectrum—some are less than 10-trillionth of a meter long. These extremely powerful energy waves can penetrate most substances. Even a brief exposure to gamma rays is deadly for most living things.

X rays were given their name because scientists were mystified by these powerful energy waves when they were discovered in the 1890s ("X" stands for "unknown"). Today the penetrating power of X rays is used in medicine to produce images of human bones and other internal structures on photographic film.

Ultraviolet radiation begins where X rays end and extends all the way to violet light, the shortest wavelength of visible light. Although not as powerful or penetrating as X rays, ultraviolet radiation is still dangerous to living things. Only a small percentage of UV radiation coming from the sun reaches the earth's surface, but even this small amount is enough to burn human skin and to kill small organisms such as plankton.

Visible light is the part of the electromagnetic spectrum containing wavelengths to which the eyes of humans and other animals are sensitive. Sunlight looks white but is actually made up of all the different wavelengths of the visible light spectrum, from short-wave violet light to long-wave red light. A red object appears red because it reflects the wavelengths of red light and absorbs the other wavelengths of the visible spectrum.

High Energy

Wavelength in micrometers (μm)

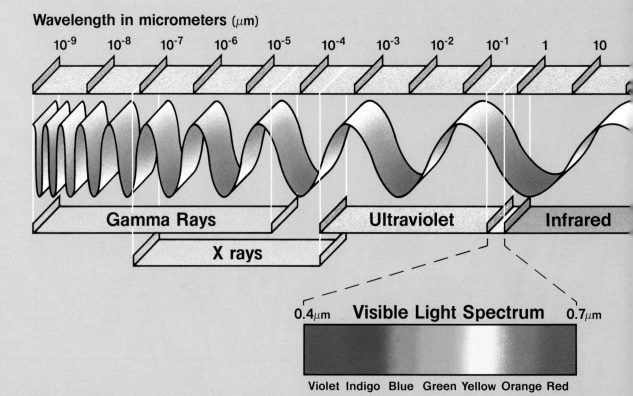

10^{-9} 10^{-8} 10^{-7} 10^{-6} 10^{-5} 10^{-4} 10^{-3} 10^{-2} 10^{-1} 1 10

Gamma Rays

Ultraviolet

Infrared

X rays

0.4μm **Visible Light Spectrum** 0.7μm

Violet Indigo Blue Green Yellow Orange Red

Infrared radiation, lying just beyond the red end of the visible light spectrum, makes up about 60 percent of the energy given off by the sun. Humans and most other animals cannot see infrared radiation, although we can feel it as heat. Snakes have special sense organs that detect the small amounts of infrared radiation given off by warm-blooded animals. Camera film sensitive to infrared makes it possible to take photographs in places where there is no visible light.

Microwaves are very high frequency radio waves, having the shortest wavelengths of all radio waves. Microwaves were discovered in the late 1800s, but devices that could produce these energy waves did not come into use until World War II. Now microwaves play a major role in everyday life. They are used in radar, in long-distance communications, in medicine, and, of course, in microwave ovens.

Radio waves cover the greatest area on the electromagnetic spectrum. Some types have wavelengths of only a few centimeters, but at the far end of the scale, there are radio waves with wavelengths many kilometers long. The different kinds of radio waves are used primarily to carry information over great distances. Some wavelengths are used for AM/FM radio broadcasts, others for television, and still others for short-wave, mobile, and citizen-band radio.

Low Energy ⟶

10^2 10^3 10^4 10^5 10^6 10^7 10^8 10^9 10^{10} 10^{11} 10^{12} 10^{13}

Radio Waves ⟶

Microwaves

Energy produced by our sun travels through space in the form of electromagnetic radiation. This energy travels at the speed of light — 299,792 kilometers (186,282 miles) per second. Different "kinds" of electromagnetic radiation are defined by their wavelengths and the amount of energy they contain. Together, all the different wavelengths, from short-wave gamma rays to long-wave radio waves, make up the electromagnetic spectrum.

wavelengths. All the other kinds of electromagnetic radiation, each with its own characteristic wavelength, fall in between these two extremes.

The amount of energy that a particular electromagnetic wave has depends upon its wavelength. The shorter a wave's length, the more energy that wave has. Gamma rays, for example, with the shortest wavelengths in the electromagnetic spectrum, have the greatest amount of energy. Radio waves, with the longest wavelengths, have the least.

Roughly 95 percent of the solar energy that strikes the earth is in the region of the spectrum that includes visible light, infrared radiation, and ultraviolet radiation. Visible light is "visible" because we can see it; that is, our eyes are sensitive to the wavelengths that make up this kind of energy. Infrared radiation, on the other hand, is invisible. But we can feel this type of energy as heat. **Ultraviolet (UV) radiation** is also invisible. Waves (or rays) of UV radiation have shorter wavelengths than either visible light or infrared. Thus, UV radiation is the most powerful of these three types of solar energy.

It is UV radiation coming from the sun that drives the ozone cycle in the stratosphere. The ozone-making process begins high up in the stratosphere, where powerful UV rays coming from the sun strike oxygen molecules (O_2). When an oxygen molecule is hit by a high-energy UV ray, the oxygen molecule absorbs the ray's energy. As a result, the bond that holds the two oxygen atoms together breaks, and the oxygen molecule splits apart into two single oxygen atoms.

$$O_2 \rightarrow O + O.$$

The individual oxygen atoms (O) quickly join with nearby oxygen molecules (O_2) to form ozone (O_3).

$$O + O_2 = O_3.$$

At the same time that oxygen molecules are being split apart, however, ozone molecules are also being hit by UV rays. When they are, they also absorb the rays' energy and break apart, leaving behind oxygen molecules and single oxygen atoms.

$$O_3 \rightarrow O_2 + O.$$

You can probably guess what happens next. Single oxygen atoms quickly combine with other oxygen molecules, forming new ozone molecules, which in turn are broken down to start the whole process again.

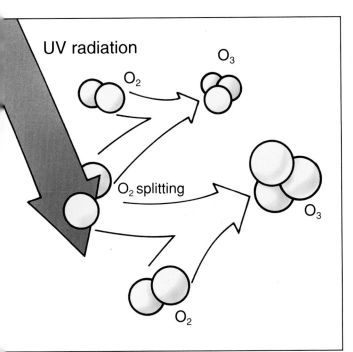
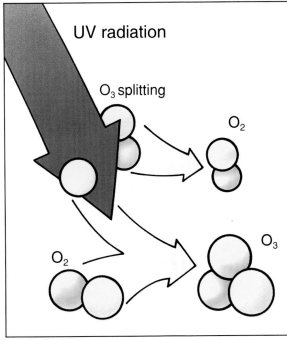

In the stratosphere, oxygen molecules that absorb high-energy UV rays split apart. Each of the oxygen atoms released in the process can join with O_2 to form O_3. When an ozone molecule absorbs UV rays, it breaks apart, leaving an oxygen molecule and a single oxygen atom, which can go on to join another O_2 and form O_3 once again.

Ozone in the Wrong Place

Ninety percent of the ozone in the atmosphere is in the stratosphere, where it protects us from harmful UV radiation. The remaining 10 percent is in the troposphere, the region of the earth's atmosphere closest to the planet's surface. Ozone in the troposphere doesn't protect living things from UV rays. Instead, it is a form of pollution, part of smog. It forms when sunlight interacts with car exhaust and industrial chemicals in the air. Unfortunately the ozone produced in the troposphere does not float up to join the helpful ozone in the stratosphere. It just collects at ground level and causes all kinds of problems.

Ozone in the troposphere is not just annoying. It is poisonous to plants, and it damages the lungs of animals, including people, that breathe too much of it. Breathing in a lot of ozone may also interfere with the body's ability to fight off infection.

Ozone pollution is a common problem in large cities. Hundreds of thousands of cars, trucks, buses, and other vehicles pump huge amounts of exhaust into the air each day. Bring all that exhaust together with sunlight, and you have a serious ozone problem.

In some cities, such as Los Angeles, the problem is so severe that people are taking steps to clear the air of as much ozone and other kinds of pollution as possible. Products like charcoal starter fluid, which contain chemicals that add to the tropospheric ozone problem, are being banned. Drive-through lanes at fast-food restaurants may

The smog that hangs over large cities such as Los Angeles contains ozone. When near the earth's surface like this, ozone is a harmful pollutant in the air.

be closed so people won't be letting their cars idle while waiting for a hamburger and fries. Wood-burning fireplaces may become relics of the past, along with gasoline-powered lawn mowers. People hope that these and other small steps will help reduce the amount of ozone there is in the air around large metropolitan areas.

As a result of this ongoing ozone cycle, about the same amount of ozone is produced as is broken down in the stratosphere. Therefore, the amount of ozone gas in the stratosphere normally stays about the same. In other words, the thickness of the ozone layer remains fairly constant.

Our Ozone Shield

A constant and stable ozone layer is important for life on earth because the high-energy UV rays that are absorbed in the ozone layer are extremely dangerous. UV radiation kills some types of living things and can damage most others. Expose certain types of bacteria to high-energy UV rays, for example, and they die within seconds. Plants, both on land and in the oceans, can be severely damaged or destroyed by UV rays. When animals, including people, are exposed to UV radiation, the powerful rays can burn skin, damage eyes, and bring about harmful, permanent changes in cells that can lead to skin cancer and other problems.

By absorbing most of the UV radiation that comes from the sun, ozone molecules in the ozone layer form a sort of shield that protects life on earth from harmful, even deadly, UV rays. Without the ozone layer, few living things would survive on our planet.

This ozone shield—the earth's armor against UV radiation—first formed in the stratosphere more than 500 million years ago. It was because of the ozone layer that plants and animals were able to spread across the face of the earth, from pole to pole. The ozone layer shielded our earliest human ancestors from harmful W rays. Today it protects us along with all other life on earth. The possibility of thc ozone layer being destroyed or even damaged is frightening. But that is what is happening today.

The first sign that the earth's ozone shield was in trouble came

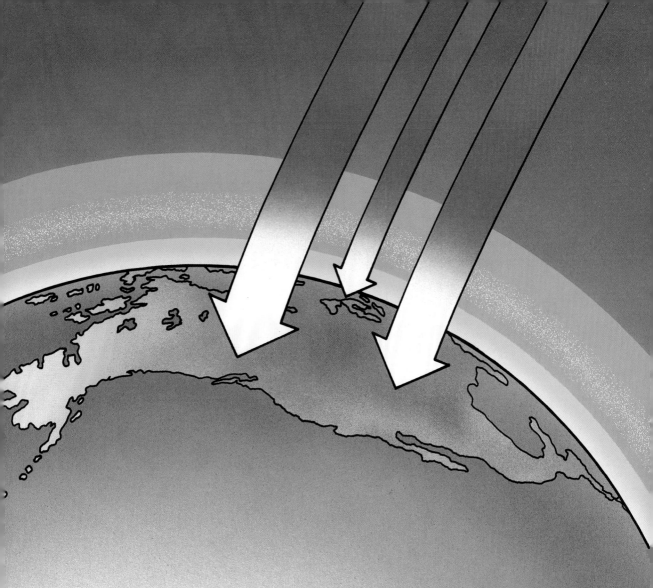

The ozone layer acts as a shield by absorbing harmful UV rays and preventing them from reaching the earth's surface.

from a most unexpected place. In the 1980s, British scientists working in Antarctica discovered that every spring ozone was disappearing from the stratosphere above them. They had found a "hole" in the ozone layer.

Large-scale stratospheric ozone depletion was first discovered over Antarctica, the frozen continent at the bottom of the world.

CHAPTER

2

Trouble at the Bottom of the World

In 1956, a Dobson spectrophotometer was shipped down to Halley Bay, a British Antarctic research station hundreds of kilometers across Antarctica from McMurdo. The Dobson was used then like it is today—to measure the amounts of ozone and other substances in the atmosphere.

Using their Dobson, scientists working at Halley Bay carefully kept track of what was happening in the ozone layer above Antarctica. For roughly 15 years, they found that the amount of ozone in the stratosphere overhead stayed fairly constant from year to year. In the late 1970s, however, they began to notice a change. Every spring—that's September to November in Antarctica— the amount of ozone would decrease quite suddenly. Ozone levels would remain low until early November, when they would begin to gradually increase. By late November, the amount of ozone in the stratosphere over Halley Bay would be pretty much back to normal.

This temporary springtime loss of ozone was troubling, and a bit of a mystery. For several years, the British scientists simply kept an eye on the situation and continued to monitor ozone in the stratosphere overhead.

By the spring of 1982, this seasonal ozone loss was much worse than ever before. The Dobson readings were so low, in fact, that the British scientists began to doubt that their data could be correct. They knew that it is fairly common for equipment to fail or behave oddly in Antarctica's extreme cold. Thinking there might be something wrong with their Dobson spectrophotometer, the researchers had a new one shipped down to Halley Bay. When it arrived, they set it up next to the old one and let the two machines take ozone measurements together.

It wasn't long before the scientists realized that there hadn't been anything wrong with the old Dobson spectrophotometer because both instruments gave the same readings. The problem was in the ozone layer. For some reason, huge amounts of ozone were disappearing from the stratosphere over Antarctica each spring. As ozone mysteriously disappeared, the ozone layer above the continent became thinner. Simply put, each spring a sort of hole formed in the ozone layer at the bottom of the world.

In 1985, the British scientists finally told the world about the ozone hole they had discovered over Antarctica. Their findings caught scientists in other countries by surprise. Researchers at the National Aeronautics and Space Administration (NASA) in the United States, for example, had been using a Nimbus 7 satellite since 1978 to keep an eye on the earth's stratospheric ozone layer from space. On board the satellite was a Total Ozone Mapping Spectrometer (TOMS), an instrument that can measure ozone levels all around the world. When NASA scientists looked back at their data from earlier years, it turned out that the TOMS *had* reported large losses of ozone over Antarctica each spring. But the data had seemed so strange that the researchers assumed that the TOMS needed adjustment and so ignored the low readings! However, after looking at the data again and doing some

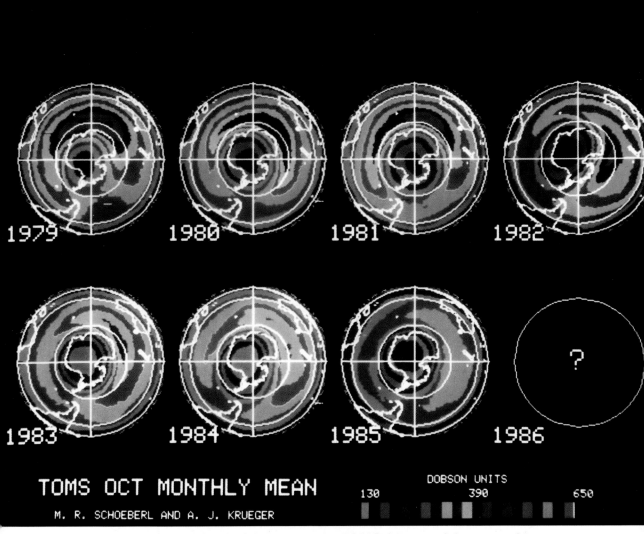

TOMS OCT MONTHLY MEAN

M. R. SCHOEBERL AND A. J. KRUEGER

DOBSON UNITS

130   390   650

Total Ozone Mapping Spectrometer (TOMS) images of the atmosphere above Antarctica from 1979 to 1985. Areas of the greatest ozone loss are shown in purple.

recalculating, NASA scientists realized that the British researchers were indeed correct about Antarctica's ozone loss. Several other research groups, using different ozone-measuring methods, confirmed the findings, too. There was no doubt about it—a hole did form each spring in the ozone layer at the bottom of the world.

And the hole was big. Pictures sent from space by the TOMS showed that when the hole was at its worst—from late September to early October—it extended over most of the Antarctic continent. The hole covered an area of nearly 20 million square kilometers (about 8 million square miles)—an area larger than the United States. Most of the ozone depletion was taking place in the lower half of the stratosphere, at an altitude between roughly 14 to 24 kilometers (9 to 15 miles) above the earth's surface.

A huge loss of ozone in the ozone layer was not something scientists had expected. More information about this strange "hole in the sky" was urgently needed.

During the Antarctic spring of 1986, Susan Solomon, an atmospheric scientist with NOAA, led a team of researchers from the United States on the first National Ozone Expedition (NOZE). Working at McMurdo, NOZE scientists used a variety of ground-based instruments to measure ozone and other molecules in the stratosphere over Antarctica. They discovered that the amount of stratospheric ozone had decreased dramatically once again. Interestingly, their instruments also detected large amounts of chlorine-containing substances in the region of the hole.

Chlorine (Cl) is a strong-smelling gas. You have probably smelled chlorine before, especially if you do a lot of swimming in pools. The strange, sharp odor of the water in public swimming pools comes from the chlorine that is added to the water to kill bacteria. In many places, small amounts of chlorine are

Dr. Ryan Sanders, a member of the 1986 National Ozone Expedition (NOZE) team, struggles against fierce winds at McMurdo.

also added to drinking water for the same purpose. Chlorine atoms are extremely reactive—they can rapidly combine with other molecules or atoms around them. When chlorine combines with some kinds of molecules, the molecules change as a result. In some cases, they break apart and are destroyed.

The next step in studying the Antarctic ozone hole was to measure ozone and these chlorine-containing compounds in the stratosphere directly. In the spring of 1987, more than 150 scientists and support personnel representing 19 research organizations from four countries joined forces to carry out the Airborne Antarctic Ozone Experiment. In addition to ground- and satellite-based monitoring devices, this research effort also involved using airplanes to fly into the ozone hole above Antarctica. Two very special planes were used. One was a DC-8 packed with scientific equipment and sampling devices. The other was a high-flying military U-2 spy plane that was converted into a similarly equipped research aircraft and renamed the ER-2.

These two airborne laboratories flew 12 missions over Antarctica. The planes took off from the tip of South America and flew south across the frigid continent of Antarctica into the hole. These were dangerous missions. Mechanical problems probably would have meant a fatal crash landing on Antarctica's frozen surface. The air temperature outside the planes was –90°C (–130°F), and the pilots

While flying through Antarctica's ozone hole aboard the specially equipped DC-8, a researcher uses instruments to gather data about ozone outside the plane.

34

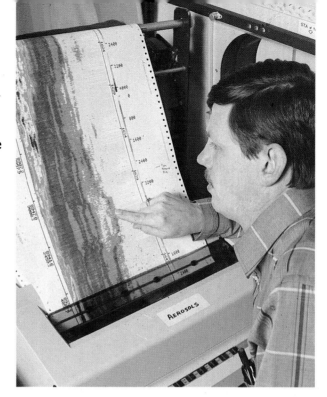

Another researcher examines data as the plane flies through the ozone hole.

had to battle winds of 150 knots (170 mph). A hurricane has winds of about 80 knots, so you can begin to imagine how rough those plane rides must have been!

As the pilots struggled to keep the planes on course above Antarctica, the scientists aboard used their instruments to sample the air outside. What did they find? As the planes flew into the hole, the amount of ozone in the surrounding air decreased sharply. But the amount of a particular chlorine-containing substance called **chlorine monoxide (ClO)** increased dramatically. There was more than 100 times more ClO inside the hole than there was outside.

All the evidence now pointed to chlorine-containing compounds as being the ozone destroyers high above Antarctica. And most of the researchers were fairly certain that the source of this chlorine in the stratosphere was a group of chemicals called chlorofluorocarbons.

35

Products that were made with or contained chlorofluorocarbons (CFCs) quickly became part of everyday life in most industrialized countries.

3

CFCs and Ozone Destruction

Chlorofluorocarbons (CFCs) are human-made chemicals that were invented in the 1930s. There are many different kinds of CFCs, but they all contain the same basic elements: chlorine, fluorine, and carbon. Different CFCs contain different numbers of chlorine, fluorine, and carbon atoms. Each CFC is identified by a number. For example, CFC-11, CFC-12, and CFC-113 are three types of CFCs that have been widely used.

Chlorofluorocarbon molecules are very stable chemicals. That means they don't easily break down or react with other substances. CFCs do not burn, are not poisonous, and do not damage the containers in which they are kept.

Being such stable chemicals, CFCs have been used for all sorts of different tasks. For example, CFC-12, which is also called Freon, became one of the most popular liquid coolants ever invented for refrigerators and air conditioners. Several other CFCs worked very well as propellants in aerosol spray cans, in manufacturing foam cushions and carpet padding, and in making plastic Styrofoam containers and packaging. Still other CFCs were used to clean delicate electronic equipment, such as computer chips and circuit boards.

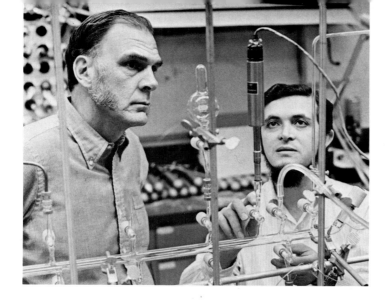

F. Sherwood Rowland (left) and Mario J. Molina

Because they worked so well and had so many uses, it wasn't long after their invention that CFCs became a part of modern life in most of the world's industrialized countries. CFCs appeared to be the perfect industrial chemicals. They caused no problems here on the earth, seeming completely safe for people and for the environment.

But long before the ozone hole was discovered, two scientists began to have second thoughts about CFCs. Were CFCs really as safe as they seemed to be? F. Sherwood Rowland and Mario J. Molina didn't think so. These two chemists wondered if CFCs were as stable high up in the atmosphere as they were down on the ground. In 1974, the two men published a scientific paper in which they outlined their concerns about CFCs.

In their paper, Rowland and Molina explained how they thought that CFCs could damage the ozone layer. Chlorofluorocarbons are liquids that evaporate very quickly—they rapidly change from a liquid to a lighter-than-air gas. When CFCs are released into the environment down here on earth, they evaporate and get into the atmosphere.

Since CFCs are so stable, Rowland and Molina reasoned, they wouldn't combine with other molecules in the air, and so they

38

wouldn't be involved in the natural processes that remove most foreign chemicals from the lower region of the atmosphere. Instead, they would remain in the atmosphere for a long time, gradually rising through the troposphere until they reached the stratosphere.

In the stratosphere, CFCs would encounter powerful UV radiation. And while CFCs are stable at the earth's surface and in the troposphere, the bonds that hold the atoms in these chlorine-containing compounds together can be broken by powerful UV rays. When a molecule of a CFC breaks apart, chlorine atoms (Cl) are released.

Unlike CFCs, individual chlorine atoms are very reactive. Rowland and Molina knew from laboratory experiments that chlorine atoms react with ozone molecules in a way that destroys ozone. What would happen, the two scientists asked, when CFCs that had been released into the environment on earth were broken apart by UV radiation high above the earth's surface? Would the chlorine atoms released in the process destroy stratospheric ozone molecules? Rowland and Molina hypothesized that CFCs would indeed harm the ozone layer in this way.

As a result of Rowland and Molina's warning about CFCs, some steps were taken in the late 1970s to cut down on the use of these chemicals. In the United States, for example, manufacturers stopped using CFCs as propellants in some types of aerosol products. But CFCs were still used in many other ways. What's more, new types of CFCs were invented. With each passing year, the amount of CFCs released into the environment increased tremendously. At the same time, the worries about CFCs and ozone depletion were all but forgotten.

The discovery in the 1980s of a hole in the ozone layer over Antarctica changed all that. Suddenly there was strong evidence that Rowland and Molina had been right about CFCs all along. Today scientists are certain that CFCs are damaging the ozone layer just as the two chemists predicted they would.

How CFCs Destroy Ozone

When CFCs are released into the environment at the earth's surface, CFC molecules mix into the air. Most CFCs are released in industrialized nations in the Northern Hemisphere, but winds quickly distribute the chemicals throughout the troposphere worldwide. Once airborne, CFCs begin a slow journey up through the earth's atmosphere. It may take as long as 100 years, but eventually these chlorine-containing molecules reach the stratosphere.

When CFC molecules reach the upper half of the stratosphere, roughly 25 to 30 kilometers (16 to 19 miles) above the earth's surface, they begin to encounter high-energy UV radiation coming from the sun. As CFCs are bombarded by these intensely powerful UV rays, bonds between atoms in the once-stable molecules break, and chlorine atoms are released. Chlorine atoms don't remain "single" for long, however. Because they are so very reactive, they rapidly join nearby molecules. Since these reactions are taking place in the ozone layer, many of these nearby molecules are ozone molecules.

When a chlorine atom and an ozone molecule come together, the chlorine atom binds to one of the oxygen atoms in the ozone molecule, stealing it from O_3. As a result of this reaction, the ozone molecule is destroyed. What's left is a molecule of oxygen gas (O_2) and a molecule of chlorine monoxide (ClO). Remember that the scientists aboard the ER-2 flights discovered lots of chlorine monoxide when they flew into the ozone hole.

But the ozone-destroying process doesn't end there. Each ClO molecule goes on to react with other molecules nearby. When two ClO molecules come together, they combine, but only briefly. Very quickly this bigger molecule breaks apart, and in the end, oxygen gas (O_2) and chlorine atoms (Cl) are left. These chlorine atoms are now free again and ready to attack other ozone molecules. And so the reaction continues, over and over and over again.

Now you can see why CFCs are so dangerous. When CFCs enter the atmosphere on earth, they slowly but surely transport the chlorine they contain up into the stratosphere. There, once released from CFC molecules, free chlorine atoms can destroy massive amounts of ozone. Atmospheric scientists estimate that on the average, a single atom of chlorine has the potential to destroy 100,000 stratospheric ozone molecules. And by 1992, more than 20 million

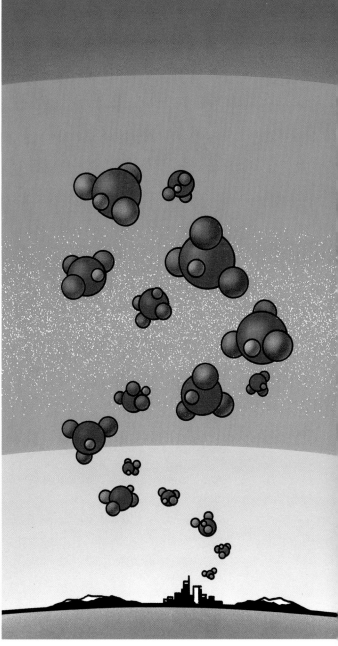

Chlorofluorocarbons gradually rise up through the atmosphere until they reach the stratospheric ozone layer.

41

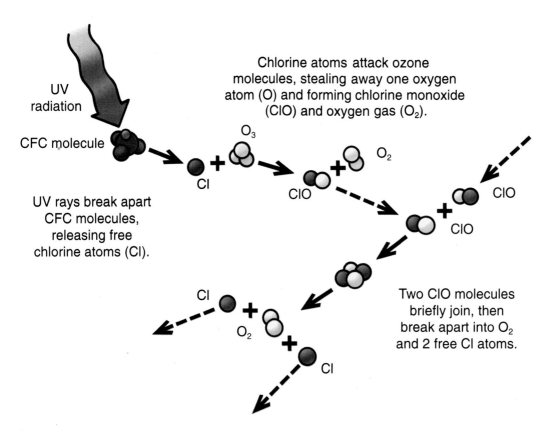

UV radiation

CFC molecule

UV rays break apart CFC molecules, releasing free chlorine atoms (Cl).

Cl

O_3

Chlorine atoms attack ozone molecules, stealing away one oxygen atom (O) and forming chlorine monoxide (ClO) and oxygen gas (O_2).

O_2

ClO

ClO

ClO

Cl

O_2

Cl

Two ClO molecules briefly join, then break apart into O_2 and 2 free Cl atoms.

Chlorine atoms destroy ozone by removing oxygen atoms from ozone molecules. Each chlorine atom has the potential to destroy many thousands of ozone molecules in this way.

tons of CFCs had been pumped into the atmosphere. That means that an almost incomprehensible number of chlorine atoms—trillions of them—are now in or on their way up to the stratosphere.

Chlorine Reservoirs

So why hasn't the ozone layer been completely destroyed by all the chlorine already transported up to the stratosphere by CFC molecules? First of all, remember that ozone is continually being formed by way of the ozone cycle, so some of the ozone that chlorine destroys is naturally replaced. Furthermore, there are certain types of molecules in the stratosphere that can bind up chlorine atoms and stop their attack on ozone. These ozone-protecting molecules are called chlorine **reservoir molecules.** As long as chlorine atoms are tied up as part of these reservoir molecules, they cannot attack ozone. Some of the most important chlorine reservoirs in the stratosphere are molecules that contain the element nitrogen.

Reservoir molecules can actually remove chlorine from the ozone layer. Some, but not all, chlorine reservoir molecules are heavy enough that they gradually drift down from the stratosphere into the troposphere. Once in the troposphere, rain and snow wash the molecules—and the chlorine atoms they contain—out of the air and down to the earth's surface.

But the ozone cycle and reservoir molecules can only do so much. Years ago, Rowland and Molina recognized that if the world kept releasing CFCs into the environment, eventually the natural systems helping to maintain and protect the ozone layer would be overwhelmed. In time, there would be so much chlorine in the stratosphere that ozone would be destroyed faster than it could be replaced. Once that happened, the two scientists predicted, there would be a slow thinning of the ozone layer all around the world.

A very gradual thinning of the earth's ozone layer has, in fact, taken place. Over the past 10 years or so, the amount of ozone in the stratosphere worldwide has decreased. But no one, not even Rowland and Molina, predicted an ozone hole over Antarctica.

Why is it there? Why is so much ozone destroyed so rapidly in the stratosphere above the Antarctic continent that each spring a hole forms in the ozone layer? It has taken years of research in the coldest place on earth to answer that very important question.

Shoulder to shoulder and knee to knee, scientists and support personnel endure the long, cold flight to McMurdo Station aboard a military cargo plane.

44

CHAPTER

4

Antarctica's Ozone Hole

Imagine being crammed together with several dozen other people, sitting awkwardly in uncomfortable webbed seats inside the hold of a cold, noisy military cargo plane. An industrial-strength seat belt keeps you in place whenever the plane swerves and lurches as it is tossed around by the winds outside. You are wearing the foam rubber earplugs you were given when you climbed aboard, but the roar of the engines is still deafening.

Everyone around you is dressed in exactly the same outfit: a big red down parka with a fur-trimmed hood, black pants with lots of pockets, red and black wool shirt (with long underwear peeking out the neck), and huge white boots that look like something Mickey Mouse would wear. It's all part of the polar survival gear that you will wear whenever you're outside in the weeks to come.

The plane has been airborne for five hours. By now your feet and hands are icy cold, and your legs are cramped from sitting still for so long. You've eaten all the sandwiches, fruit, and cookies that were in the box lunch you were handed as you boarded the plane. But you still have several hours yet to go before you reach your destination: McMurdo Station on Ross Island, Antarctica.

There are no windows to look out of on this flight; only the pilots get a view. But even if there were windows, there would be little to see—just stars overhead and darkness down below. It is August 22 and still winter at the bottom of the world.

Even at the best time of year, during the daylight summer season, traveling to Antarctica is a long, hard, dangerous journey. Most flights carrying American researchers down to "the Ice"— a nickname for Antarctica—depart from Christchurch, New Zealand. The distance between Christchurch and McMurdo Station is more than 5,000 kilometers (3,100 miles).

The flight is not only difficult but also unpredictable. Weather conditions in and around Antarctica change remarkably fast. Blizzards can spring up in seconds, reducing visibility on the ground to zero. Frequently, when a plane loaded with supplies and scientists is already hours into the flight, the weather at McMurdo will suddenly change for the worse. When that happens, the pilot has no choice but to turn the plane around and fly all the way back to New Zealand. Nearly all experienced Antarctic researchers have been on at least one of these "boomerang" flights.

But this particular plane is traveling to Antarctica at one of the worst times of the year. In August, temperatures at McMurdo average −30°C (−22°F), and the winds can be ferocious. Swooping down out of the sky to set down on Antarctica's dark, frozen landscape at this time of year is a challenge for the pilot and a heart-stopping experience for the passengers. Why would these people want to risk flying to Antarctica at this time of year? Because it's the best time to study the Antarctic ozone hole.

This flight is part of WINFLY, the "winter fly-in" program of the U.S. Antarctic Research Program. Before WINFLY was established, there were rarely any flights into or out of McMurdo at this time of year. But to study ozone destruction over Antarctica, it is necessary to get ozone researchers there before the hole

Antarctica's unpredicatable weather makes flying into and out of McMurdo difficult and often very dangerous. Notice that this plane is equipped with skis for landing on the continent's frozen surface.

begins to form, so the entire process—from start to finish—can be studied firsthand.

In the years since the discovery of Antarctica's ozone hole, our understanding of the problem has grown considerably. Programs such as WINFLY have given ozone researchers new insights into stratospheric ozone depletion. Scientists now know that several special characteristics of the Antarctic environment are the key to why a hole forms in the ozone layer there. These factors, working together, lead to the rapid destruction of massive amounts of stratospheric ozone each spring.

The Coldest, Windiest Place on Earth

On the edge of McMurdo, two WINFLY scientists struggle to get a small weather balloon ready to launch. In just the last few minutes, the wind has died down enough to make a launch possible, but the intense cold makes every job difficult. One of the researchers slips off warm mittens to attach an instrument that measures temperature and wind speed to the balloon. By the time the instrument is attached, the researcher's fingers are numb and white.

When the balloon is released, it floats up through the dark sky, through the troposphere, and into the stratosphere. At about 20 kilometers (12.5 miles) up—close to the middle of the ozone layer—the balloon's instrument records an air temperature of –78°C (–108°F), which is not an unusual reading for this time of year. In winter, the stratosphere above the Antarctic continent gets colder than it does anywhere else on earth. Temperatures frequently drop below –80°C (–112°F).

Antarctica is also one of the windiest places on earth. In May and June, strong winds in the stratosphere begin to blow clockwise around the continent. These howling stratospheric winds gradually form an enormous ring of moving air, called the **Antarctic polar vortex,** that swirls around and around, far above the frozen land.

The mass of air inside the vortex, however, is quite still. As long as the vortex winds continue to blow, the air inside the vortex doesn't mix with warmer air outside the vortex. As the Antarctic winter progresses, the air trapped within the vortex gets steadily colder. Eventually, the temperature of that air falls below a critical point, and rare clouds appear in the sky over Antarctica.

48

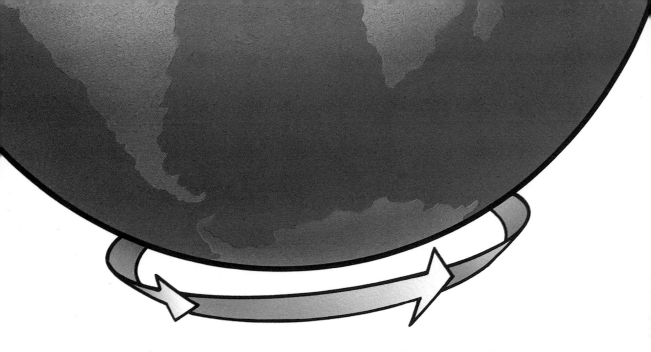

During winter, the Antarctic polar vortex forms in the stratosphere above Antarctica, isolating a mass of air over the continent that gradually becomes colder and colder.

Polar Stratospheric Clouds

Bundled up in their red polar parkas, three scientists stand huddled over a compact-looking device set up in the center of a small metal building on the outskirts of McMurdo. The instrument is called a **lidar,** which is short for "light radar." Directly above the lidar, a panel in the building's flat roof has been slid open to reveal the twilight sky. Frigid air pours down through the opening in the roof, but the researchers don't seem to notice. Like other WINFLY scientists, these researchers are in McMurdo during the last days of August to monitor what is happening in the atmosphere overhead. With the lidar, they are using laser beams to study unusual clouds that form inside the Antarctic polar vortex.

Clouds rarely, if ever, form in the stratosphere above most parts of the world. But the situation over Antarctica is different. During the winter, temperatures inside the Antarctic polar vortex fall so

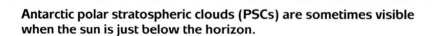

Antarctic polar stratospheric clouds (PSCs) are sometimes visible when the sun is just below the horizon.

low that water vapor and several other types of molecules in the stratosphere condense into extremely small icy particles. These icy particles, in turn, make up **polar stratospheric clouds (PSCs).**

Polar stratospheric clouds are quite different from the types of clouds we frequently see in the troposphere. For one thing, PSCs can be very large—a single cloud may be more than two kilometers thick and hundreds of kilometers long. Despite their huge size, PSCs are nearly invisible from the ground. But with the help of a lidar, scientists can "see" PSCs. The instrument works by sending rapid pulses of bright green laser light straight up into the stratosphere overhead. When the laser beams collide with the icy particles that make up PSCs, some of the light bounces off the particles and is reflected back toward earth. Using a telescope that is lined up with the laser, researchers measure the light that bounces back with each pulse. From the brightness of the reflected light they can tell how many particles there are in the PSCs overhead. By recording the time it takes for the light to return, they can calculate how far above the ground the clouds are.

Atmospheric scientists discovered PSCs many years ago. Above Antarctica, these unusual clouds tend to form in roughly the same part of the stratosphere in which the ozone layer is found. Up until the appearance of the ozone hole, however, PSCs were thought to be little more than a curious wintertime feature in the Antarctic sky. But it turns out that PSCs work together with two other factors—UV radiation and chlorine compounds—to bring about ozone destruction over the continent each spring.

How the Ozone Hole Forms

When the sun sets in the Antarctic around the end of March each year, its disappearance marks the beginning of a long, dark winter. Once the last rays of sunlight have faded away, temperatures on land and in the air fall very quickly.

In the stratosphere, high-altitude winds that create the polar vortex begin to blow around the continent. Isolated from warmer air outside the vortex, the air inside gets colder and colder. Eventually, it is cold enough for PSCs to form. And that is when the trouble really begins.

Drifting around inside the vortex are reservoir molecules that have bonded with chlorine atoms and in doing so prevented them—so far—from attacking ozone. When PSCs form above Antarctica, chlorine reservoir molecules bind to the icy particles that make up the clouds. Once this happens, complex chemical reactions begin to take place that result in molecules of chlorine gas (Cl_2) being released from the reservoirs. In this form, however, chlorine doesn't attack ozone. It just collects inside the vortex.

All through the long, dark winter, especially during July and August, the chemical reactions taking place on the surfaces of the PSC particles continue, and more and more Cl_2 builds up inside the vortex. At this point, the stage is set for ozone destruction. All that is needed is a trigger to get the process going.

That trigger comes in late August, when the sun begins to rise. As the first rays of spring sunlight strike the stratosphere high over the frozen continent, conditions change very rapidly. The UV rays coming from the sun strike the Cl_2 molecules inside the vortex. The molecules break apart, releasing billions of chlorine atoms that begin an attack on ozone molecules. The result is massive ozone destruction. Before long, so much ozone is destroyed inside the vortex that an ozone hole is formed.

Ozone destruction continues—and the hole remains—until conditions in the stratosphere above Antarctica change. This change usually begins in early October, when the continent and the air above it finally begin to warm up. Warmer temperatures in the stratosphere melt the icy particles that make up PSCs. The PSCs disappear, and the reservoir molecules that were bound to the icy particles are released. Free at last, the reservoir molecules bind Cl atoms once again, and ozone destruction stops.

By early November, the strong stratospheric winds circling Antarctica die down, and the polar vortex breaks up. As it does, ozone-rich air from outside the vortex flows in, and much of the ozone that was destroyed is replaced. In a sense, the hole in the ozone layer fills in. Usually by the end of November, the amount of ozone in the stratosphere over Antarctica has almost returned to normal. The next winter, however, the cycle will begin again.

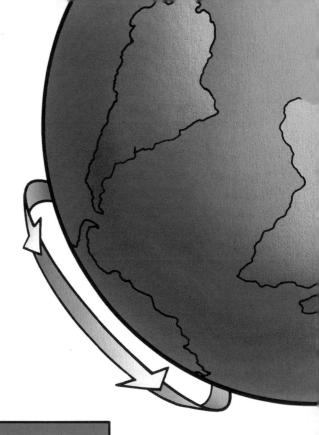

In the stratosphere, chlorine from CFCs binds to reservoir molecules.

During the Antarctic winter, PSCs form inside the polar vortex.

Chlorine reservoir molecules bind to the icy particles in the PSCs.

As a result of this binding process, Cl_2 is released. Throughout the winter, Cl_2 accumulates inside the vortex.

At sunrise, UV rays split Cl_2 molecules, releasing enormous numbers of free Cl atoms, which rapidly destroy massive amounts of ozone.

A summary of the steps that lead to the formation of an ozone hole above Antarctica each spring.

Measuring Ozone Loss

Figuring out how and why a hole forms in the ozone layer above Antarctica has been a tremendous challenge for scientists. So has trying to measure how much ozone is destroyed each spring. The ozone hole and the conditions that cause it are constantly changing. Imagine trying to measure something that is different from moment to moment!

There are several ways to measure ozone in the stratosphere. Some researchers use ground-based instruments that look up into the atmosphere and measure ozone levels overhead. The Dobson spectrophotometer is one of these types of instruments. In addition to measuring ozone, a Dobson spectrophotometer can be used to measure chlorine monoxide and other types of molecules that are involved in the ozone destruction process.

A Dobson spectrophotometer measures how much of a particular substance, such as ozone, there is in a thin column of air directly over the machine. It gives those measurements in Dobson Units (DUs). When there is a normal amount of ozone in the stratosphere over Antarctica, the average reading you would get using a Dobson is around 300 DUs. In the spring of 1985, the British scientists at Halley Bay were getting ozone readings of roughly 180 DUs. During the 1987 ozone hole, scientists working at several Antarctic research stations were getting readings of 120 to 125 DUs. And in October 1992, measurements taken at the South Pole showed that ozone levels in the hole had fallen to a record low of 105 DUs.

Dobson spectrophotometers measure the total amount of ozone in a column of air directly overhead. But they can't give important details about the altitude at which the greatest amount of ozone depletion is taking place. To fill in that part of the puzzle, other scientists send ozone-measuring devices called

ozonesondes up into the stratosphere attached to helium-filled balloons.

From late August through October, ozonesondes are launched at least every other day (weather permitting) from a number of Antarctic research stations. Launching an ozonesonde is not a simple job. First a large cloth is spread out on the ground. This helps protect the balloon, which is made out of extremely thin plastic, from stones and other sharp objects. On top of the cloth, the balloon is unpacked. Next, a large nozzle—attached by a hose to a

While one scientist controls the flow of helium (right), two others gently unfold the balloon as it inflates.

tank of helium—is inserted into a long filling tube that leads to the top of the balloon. Helium is a gas that is lighter than air. It takes several minutes, and several pairs of helping hands, to fill the balloon about one-third full with helium. Finally, a small

When the balloon is about one-third full (as this one is), the filling tube at the top is tied off, and the ozonesonde is attached to the bottom of the balloon.

The balloon is launched.

Styrofoam-covered box containing the ozonesonde and an electronic transmitter is tied securely to the bottom of the balloon.

When everything is ready, the researchers wait for a break in the wind. When a moment of calm arrives, the balloon is released,

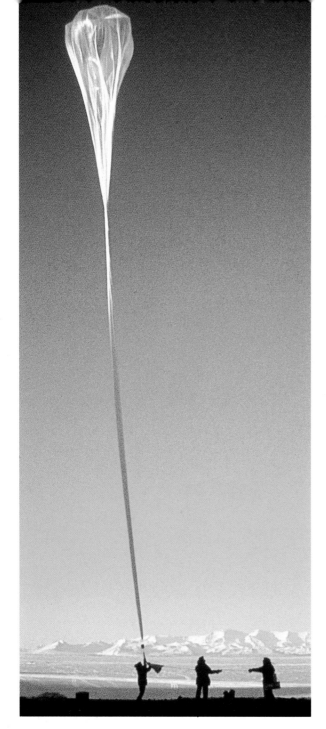

A balloon and the ozonesonde it carries begin the journey up into the Antarctic stratosphere.

and it soars up into the sky. Near the ground, the partly filled balloon looks thin and flabby. But as the balloon rises, the gas inside expands, so the balloon gets fuller and rounder the higher it goes.

As the ozonesonde travels skyward, it records the amount of ozone in the air around it. It continually sends ozone measurements back to earth. Most ozonesondes travel to the top of the ozone layer, about 35 kilometers (22 miles) above the ground. That's higher than any research airplane can fly. But by the time

Two ozone researchers battle the wind to retrieve an ozonesonde that has returned to earth not far from the research station.

the ozonesonde reaches that altitude, the helium inside the balloon's thin plastic walls has expanded to take up 200 times more space than it did at ground level. That's too much for the balloon. It bursts, and the ozonesonde falls back to earth, its job complete.

Sometimes an ozonesonde lands close enough to its launch point that someone can travel out and collect it. If it has not been damaged by its crash landing, it can be reused. More often, however, the ozonesondes land beyond the reach of scientists and disappear forever in the frozen Antarctic landscape.

Using information gathered by ozonesondes, scientists can piece together a picture of the atmosphere that shows how much ozone there is—or isn't—at various altitudes. Over the past few years, scientists have discovered that when the ozone hole is at its worst each spring, up to 98 percent of the ozone between 14 and 24 kilometers (9 and 15 miles) above Antarctica is destroyed.

Spectrophotometers and ozonesondes provide information about what is happening in very small sections of the ozone hole. Only instruments that are orbiting the earth on satellites, such as a TOMS, can give researchers a broad view of ozone destruction over Antarctica. TOMS images show how large the Antarctic ozone hole is from day to day.

When scientists bring together the information they receive from ground-based instruments, from ozonesondes, and from satellites, a fairly accurate picture of what is happening in the ozone layer above Antarctica emerges. Unfortunately, it is not a very encouraging picture. Researchers have discovered that each spring, more than half of the total amount of ozone over Antarctica is temporarily destroyed. And at certain altitudes, all of the ozone is destroyed. Furthermore, the situation is getting worse. Not only is more ozone destroyed each year, but the hole develops earlier and lasts longer.

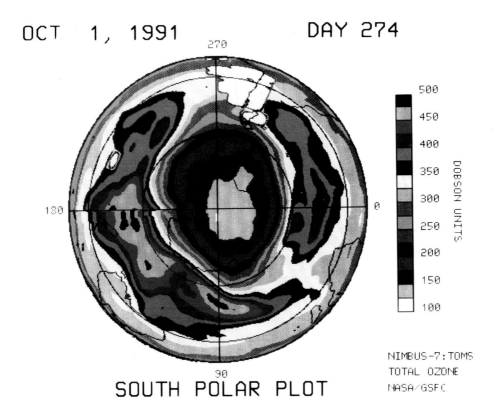

OCT 1, 1991   DAY 274

270

180   0

DOBSON UNITS

500
450
400
350
300
250
200
150
100

90

SOUTH POLAR PLOT

NIMBUS-7:TOMS
TOTAL OZONE
NASA/GSFC

A TOMS image showing ozone depletion over Antarctica on October 1, 1991. The region of greatest ozone loss is shown in lavender.

The ozone layer is earth's protective shield against harmful UV radiation coming from the sun. When the ozone hole develops at the bottom of the world each spring, there is less ozone between Antarctica and the sun. That means that more harmful UV radiation is reaching the earth's surface under the hole. The effect that this environmental change has on life on and around this ice-covered continent is what another group of Antarctic researchers is investigating.

Researchers are not yet sure how the ultraviolet radiation coming through the ozone hole each spring will affect life on and around Antarctica.

62

CHAPTER

5

Life under the Hole

The types of UV rays absorbed by the ozone layer are those that are most harmful to living things. Less ozone in the stratosphere means that more of this harmful radiation can reach the surface of our planet. Each spring, as ozone temporarily disappears over Antarctica, more UV radiation is able to pass through the atmosphere to reach the ground under the hole.

How much more? Before there was an ozone hole, Antarctica received very little, if any, UV radiation in the springtime. Since the appearance of the hole, however, the Antarctic continent and the waters around it now receive at least as much UV radiation each spring as they normally would in the middle of the summer. That is a significant change in the Antarctic environment.

For several years, biologists have been looking for evidence that this change is having an effect on plants and animals living in Antarctica. Several groups of researchers have focused their studies on minute marine plants called **phytoplankton** that live near the surface of the ocean. Phytoplankton, along with tiny marine animals called **zooplankton,** drift by the trillions through the Southern Ocean that surrounds Antarctica. Most are too small to be seen without a microscope.

Phytoplankton may be small, but they are very important. These tiny plants form the base of the Antarctic food web. Phytoplankton are eaten by zooplankton such as the shrimplike krill. Krill, and many animals that eat krill, are food for large fish, penguins, seals, and whales. No matter what the particular diet of the animals that live in the Southern Ocean, their food supply ultimately depends on the welfare of phytoplankton. Anything that affects the phytoplankton affects the other members of the Antarctic food web as well.

Getting the right amount of sunlight is critically important for phytoplankton, just as it is for all green plants. Sunlight provides the energy to drive **photosynthesis,** the process by which green plants manufacture their own food. In photosynthesis, carbon dioxide and water are combined—using energy from the sun— to form compounds like sugars and starches.

Every year, during the long period of winter darkness, a large part of the phytoplankton population in the ocean around Antarctica dies off. But with the return of spring's sunshine and

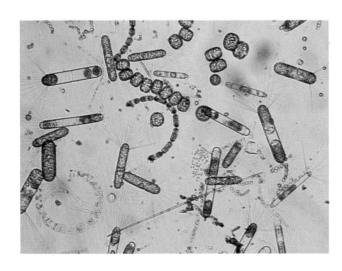

Phytoplankton are minute plants that form the base of the Antarctic food web.

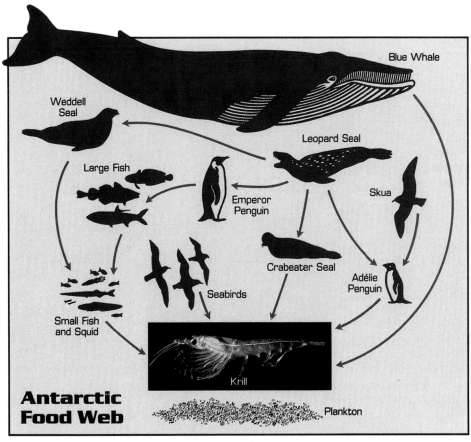

Antarctic Food Web

Blue Whale

Weddell Seal

Leopard Seal

Large Fish

Emperor Penguin

Skua

Crabeater Seal

Adélie Penguin

Seabirds

Small Fish and Squid

Krill

Plankton

If higher-than-normal amounts of UV radiation coming through the ozone hole harm plankton in the waters around Antarctica, the food supplies of all the other members of the food web could be at risk.

warmth, the phytoplankton that have survived the winter begin to divide and multiply. The population grows tremendously, so that by summer, the waters around Antarctica teem with phytoplankton and all the animals that depend on them for food.

Although phytoplankton need sunlight for photosynthesis, they can be damaged by UV radiation, especially when they are growing and dividing rapidly. Because of the ozone hole, phytoplankton

65

Palmer Station is a small U.S. research station on the Antarctic Peninsula.

in Antarctic waters are now being exposed to summertime levels of UV radiation in early spring, just as they are beginning their annual population explosion. These higher-than-normal levels of UV radiation appear to be causing problems for phytoplankton.

Palmer Station is a small U.S. research station located on a tiny island close to the western coast of the Antarctic Peninsula. Palmer is quite different from McMurdo. There are just a handful of buildings, a helicopter pad, and a dock. There are no runways for airplanes; the station can only be reached by ship. Researchers must sail down from the tip of South America, across the stormy Southern Ocean to reach this isolated outpost.

Palmer Station is in a wonderful location, however, for studying phytoplankton. Over the past several years, phytoplankton have been receiving a lot of attention at Palmer. Biologists working there have been scrutinizing these tiny marine plants, looking into how phytoplankton are affected by the higher-than-normal levels of UV radiation they have been receiving through the ozone hole each spring.

Dr. Deneb Karentz, a biologist from the University of California at San Francisco, has spent several seasons at Palmer. In her research, she works with different kinds, or species, of phytoplankton called **diatoms.** Diatoms are common in the waters around Palmer Station and are fairly easy to collect. In the station's

Dr. Deneb Karentz skis near Palmer Station.

laboratory, the tiny plants are used in a variety of experiments in which they are exposed to different levels of UV radiation using special lamps that give off UV rays.

In all living things, the genetic material—DNA—can be damaged by UV radiation. High-energy UV rays can break the bonds that hold DNA molecules together. UV rays can cause varying amounts of damage, depending on the organism. And each organism differs in its ability to repair this damage. For instance, nearly all of the species of diatoms that Deneb Karentz and her coworkers have studied in Antarctica have some natural defenses against UV radiation. Most of the tiny plants can manufacture special sunscreen chemicals that help protect them against UV rays. Many are also able to repair damaged DNA molecules, at least up to a point. However, certain species of diatoms are much better at protecting themselves, or repairing damage, than others.

Experiments performed by Deneb Karentz and other biologists at Palmer indicate that some types of phytoplankton are more likely to be damaged or killed by UV radiation than others. These results raise some troubling questions: Will the more sensitive types of phytoplankton gradually die out as a result of the higher-than-normal amounts of UV radiation coming through the hole each spring? What will happen if these species eventually disappear from Antarctic waters? There are no clear answers to these questions yet, but the research continues.

Hundreds of miles across the Antarctic continent from Palmer, biologists working at McMurdo are also investigating the effects of UV radiation on phytoplankton. Experimenting with phytoplankton in early spring is in some ways more challenging at McMurdo than at Palmer because, at that time of year, thick ice covers the sea around most of Ross Island. To collect phytoplankton for their experiments, researchers at McMurdo must either travel many kilometers to the open water at the edge of the ice pack or drill holes through the ice to reach the water below.

Dr. Patrick Neale from the University of California at Berkeley and Dr. Michael Lesser from the Bigelow Laboratory in Maine have been investigating how phytoplankton under the ice near McMurdo are affected by UV rays coming through the ozone hole. Just as UV rays can penetrate water, they can also penetrate ice. During several polar spring seasons, these two researchers have set up phytoplankton experiments on the sea ice several kilometers from McMurdo, next to a fish hut that covers a hole in the ice.

Antarctic fish huts are small—but heated—rectangular, wooden shelters with a sort of trapdoor in the floor. Each portable hut can be positioned so that the trapdoor sits directly over a hole, roughly a meter in diameter, that has been drilled through the ice to the seawater below. Drilling these holes can take quite a while, because in early spring the sea ice around Ross Island is 3 to 6 meters (about 10 to 20 feet) thick. Each hole becomes an opening through which researchers can get at the marine organisms they wish to study.

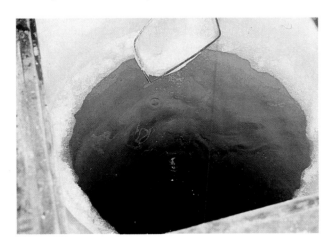

A circular hole cut into the sea ice near McMurdo Station allows scientists to reach plankton in the seawater beneath the ice.

To collect phytoplankton for their UV radiation experiments, Patrick Neale and Michael Lesser lowered a plankton net through the hole beneath their fish hut. They swirled it from side to side at the bottom of the hole so that it scraped against the underside of the ice. When they pulled the net up, it was filled with tiny bits of ice and thousands of diatoms and other phytoplankton.

The researchers transferred the phytoplankton to a boxlike experimental chamber set up outside the hut. The rectangular chamber held two seawater-filled compartments, each covered by clear Plexiglas. The compartments looked exactly the same.

Using a plankton net that can be lowered into the hole, researchers collect phytoplankton from beneath the sea ice. When the net is brought up, the tiny plants settle into the container at the net's end.

Dr. Michael Lesser (left) collects a sample of phytoplankton from one compartment of the experimental chamber set up on the sea ice.

The Plexiglas covering one compartment, however, allowed UV rays to pass through it, while the Plexiglas on the other compartment blocked UV rays.

Both compartments were kept covered by a thin black mesh cloth. The cloth cut down on the brightness of the sunlight striking the phytoplankton in the compartments. By using it, the researchers could control the light conditions in the compartments so that they were similar to the conditions these tiny plants would normally experience under the ice at this time of year.

71

Every few days, the two men would travel out to the fish hut and remove a sample of phytoplankton from each compartment. Back in a laboratory at McMurdo, the samples were analyzed to see how well the phytoplankton were photosynthesizing and growing.

The researchers soon discovered that phytoplankton taken from the compartment that was not protected from UV rays were not photosynthesizing as much and didn't grow as well as the phytoplankton on the protected side. From this and other experiments, the scientists concluded that even under the ice, higher-than-normal amounts of UV radiation affect how well some species of phytoplankton photosynthesize and grow.

The results of experiments carried out by Deneb Karentz, Patrick Neale, Michael Lesser, and other researchers show that some kinds of phytoplankton in Antarctic waters are at risk under the ozone hole. The higher levels of UV radiation present each spring can damage or inhibit the growth of certain species. These findings led researchers to conclude that, over time, the entire Antarctic phytoplankton community could change. Those species that can tolerate the extra UV radiation will probably not be affected, but those that can't probably won't do as well. The end result may be that some species flourish, while others dwindle or even die out.

No one knows exactly what such changes in the composition of the phytoplankton will mean for the Antarctic food web. There will still be trillions of tiny plants in the ocean around the continent. But will they be the "right" kinds? If it turns out that krill and other phytoplankton-eating animals in the waters around Antarctica will eat a new phytoplankton mix just as eagerly as they eat the current phytoplankton mix, there might not be a problem. But if they won't, the phytoplankton-eaters might not

Scientists are worried that harmful UV rays coming through the ozone hole may eventually damage Antarctica's fragile and unique ecosystems.

do as well as they do now, meaning there would be less food available, in turn, for the animals that eat the phytoplankton-eaters. The entire Antarctic food web could be affected.

Phytoplankton are not the only living things that are growing and developing during the Antarctic spring. This is also the time when penguins and seabirds are building nests and laying eggs.

A pair of Emperor penguins and their chick stand in the Antarctic spring sunlight.

Because of the ozone hole, many of Antarctica's animals, like this Weddell seal, are exposed to higher-than-normal amounts of harmful UV radiation each spring.

It is when mother seals come up onto the ice to give birth to their pups. Can the UV radiation coming through the ozone hole harm these Antarctic animals and their young? Fortunately, feathers and fur prevent most UV rays from damaging skin, tails, wings, and flippers. Eggs are safe because UV radiation doesn't pass through their shells. But animal eyes are sensitive and vulnerable to UV rays. No one knows what effects, if any, the higher levels of UV radiation are having on the eyes of Antarctica's animals each spring.

After several years of studying life under the Antarctic ozone hole, scientists are just beginning to understand how life on and around this snow-covered continent is being affected. But it is important that scientists continue their investigations. What they discover there may turn out to be important for life everywhere on the planet, because ozone depletion is not taking place only over Antarctica each spring. The ozone layer is being threatened worldwide.

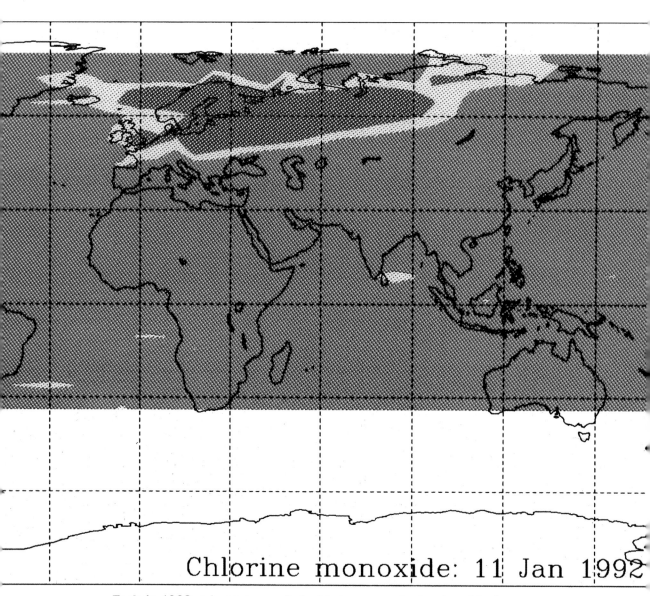

Chlorine monoxide: 11 Jan 1992

Early in 1992, a huge mass of air rich in ozone-destroying chlorine monoxide was detected over northern Europe and Russia. A similar mass appeared over North America.

CHAPTER

6

Ozone in Trouble Everywhere

In early December 1987, New Zealanders who turned on their televisions to catch the evening news were warned of a new problem facing their country. The news reports were not about higher unemployment rates or new taxes but about how to avoid being exposed to high levels of UV radiation. A huge mass of ozone-depleted air had settled over New Zealand. Until the danger passed, people were urged to stay indoors and to wear long sleeves, hats, and sunscreen if they had to be outside.

When Antarctica's ozone hole was first reported in 1985, most people—especially those living in the Northern Hemisphere—were not too worried about the problem. To them, Antarctica seemed a faraway place, frozen and isolated at the bottom of the world. The ozone depletion that took place there each spring was interesting, they thought, but not something that would affect any people except the few scientists who spent time there.

It was only two years later, in 1987, that the world learned that ozone destruction was not just a problem for penguins, phytoplankton, and Antarctic researchers. When the Antarctic ozone hole came to an end that year, something quite unexpected happened. As the polar vortex broke up, several huge masses of ozone-depleted air broke away from the stratosphere over Antarctica and drifted over other parts of the Southern Hemisphere. With the help of satellites and ground-based instruments, scientists were able to track these ozone-depleted air masses.

Some hovered above parts of the ocean. But a particularly large air mass settled over New Zealand and southern Australia. When it did, ozone levels in the stratosphere above parts of the two countries fell to record lows for that time of year. Without the normal amount of ozone in the stratosphere overhead, higher-than-normal amounts of UV radiation were able to reach the ground below. Warnings about the dangers of UV rays kept thousands of people inside their homes during the day. The ozone-depleted air mass eventually mixed with air from other parts of the stratosphere, but several weeks passed before the ozone levels returned to normal.

The escape of ozone-depleted air from Antarctica's polar vortex was rather frightening. For the first time, people realized that ozone destruction was not necessarily limited to the stratosphere above Antarctica. The event was just a glimpse of trouble ahead. Recently, scientists have discovered that stratospheric ozone depletion is much more serious—and much more widespread—than anyone had ever imagined.

Trouble at the Top of the World

Working in Antarctica, scientists had learned that at the bottom of the world, several key factors come together in just the right way for an ozone hole to form. Inside the polar vortex above this very cold place, PSCs and chlorine-containing molecules react in such a way that all it takes is sunlight to trigger massive ozone destruction.

Polar stratospheric clouds over northern Sweden

Like Antarctica, the Arctic is a very cold place. And during winter in the Northern Hemisphere, a polar vortex forms over the top of the world as well. It didn't take long after the discovery of Antarctica's ozone hole for researchers to ask a disturbing question. Could an ozone hole, like the one that forms over Antarctica each spring, form over the Arctic?

For the past few years, teams of ozone researchers from the United States, Britain, Japan, Canada, Russia, and several European countries have joined forces to study conditions in the stratosphere at the top of the world. The same two research planes that flew many missions into Antarctica's ozone hole—the ER-2 and the specially equipped DC-8—have crisscrossed Arctic skies, looking for signs of ozone destruction. Thousands of ozonesondes and other kinds of balloon-carried instruments and experiments have been launched from many places above the Arctic Circle, including northern Alaska, Canada's Northwest Territories, Greenland, Norway, and Siberia.

From space, the TOMS has provided thousands of pictures of the atmosphere above Arctic regions. In 1991, the NASA space shuttle *Discovery* carried a new satellite, the Upper Atmosphere Research Satellite (UARS), into space to help collect even more information about stratospheric ozone levels.

What have scientists learned from all of the experiments, measurements, and information-gathering instruments? They have discovered that, in some ways, conditions in the stratosphere above the Arctic are very similar to those above Antarctica. All of the factors that can lead to ozone destruction are there, including PSCs and large amounts of chlorine-containing compounds.

But researchers have discovered that some things are different at the top of the world. Temperatures in the stratosphere over Arctic regions don't get quite as cold as they do over Antarctica. And the Arctic polar vortex isn't as strong. It wobbles on its course around the North Pole, sometimes dipping down below the Arctic Circle and then swooping up again. When that happens, warmer air (with more ozone) mixes with the air in the vortex. Furthermore, the Arctic polar vortex usually breaks up earlier in the spring than does the Antarctic vortex.

Nevertheless, when the first rays of sunlight strike the stratosphere over the North Pole early each spring, some ozone destruction does take place. But for now, the destruction at the top of the world isn't as severe as it is at the bottom. Remember that more than 50 percent of Antarctica's stratospheric ozone is destroyed each spring, and at an altitude between 14 and 24 kilometers (9 and 15 miles), nearly all of the ozone is destroyed. Over Arctic regions, on the other hand, only about 15 percent of the ozone is destroyed each spring before warmer temperatures break up the polar vortex and ozone-destroying reactions stop.

A 15-percent loss may not seem like all that much compared to what happens over Antarctica. But unfortunately, most atmospheric researchers think that ozone destruction over the Arctic may be much worse in the future. The reason for this concern is that conditions in the stratosphere over northern polar regions seem to be more changeable than those over Antarctica. In the past few years, for example, springtime temperatures in the Arctic stratosphere have warmed up much earlier than expected, before large amounts of ozone were destroyed. But that may not always be the case. Sooner or later, many scientists predict, the Arctic will have a longer-lasting, colder-than-average winter. When that happens, the chances of massive ozone destruction—enough to cause an Arctic ozone hole—will be much greater.

Unexpected Trouble

Because conditions in the stratosphere over the North Pole are similar to those above Antarctica, few scientists were surprised to find ozone depletion over Arctic regions. But no one predicted what happened over the Northern Hemisphere in 1992.

The winter of 1991–1992 was an unusually warm one in the Arctic, and the polar vortex broke up very early. Only about 10 percent of the ozone above the region was destroyed that year. But in late January 1992, the ER-2 flew a series of missions over some of the most heavily populated parts of the Northern Hemisphere. Scientists on board were shocked to find high concentrations of ClO—the telltale sign that ozone was being destroyed—in the stratosphere above the United States and Canada. In fact, over eastern Canada, the plane flew through the highest concentrations of chlorine monoxide it had ever encountered, even on flights into the Antarctic ozone hole. Not

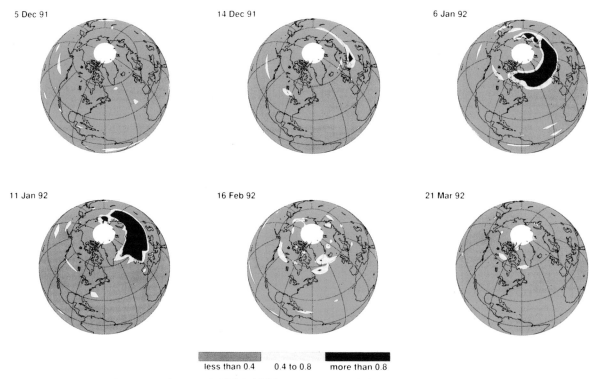

14 Dec 91

6 Jan 92

11 Jan 92

16 Feb 92

21 Mar 92

less than 0.4   0.4 to 0.8   more than 0.8

During the winter of 1991-1992, masses of ClO-rich air were detected over much of the Arctic region. Scientists are concerned that in the future, an ozone hole may form at the top of the world similar to the one that now forms over Antarctica each spring.

long after this, the earth-orbiting UARS sent back a disturbing picture of a similar ClO-rich air mass over most of Europe.

Where had all this chlorine monoxide come from? Some researchers thought it might have come from the Arctic polar vortex. They proposed that ClO-rich air broke away from the vortex and swirled southward across the globe, just like the ozone-depleted air mass that had broken away from the stratosphere over Antarctica and drifted northward over New Zealand and Australia in 1987.

Other scientists suggested that tiny dust and acid particles in the stratosphere can function like PSC ice crystals, allowing chemical reactions to occur that free ozone-destroying molecules. What makes these molecules of dust and acid so dangerous for ozone is that they are found in the stratosphere worldwide and year-round. They are primarily produced by volcanoes that erupt with enough force to shoot tons of dust, ash, and other substances all the way up to the stratosphere. There is now evidence that several recent volcanic eruptions, like that of Mount Pinatubo in the Philippines in 1991, are responsible for at least part of the stratospheric ozone depletion taking place away from the polar regions.

The amount of ozone that was temporarily destroyed over the Northern Hemisphere in the spring of 1992—5 to 8 percent— wasn't enough to cause an ozone hole to form. But that ozone loss occurred above some of the most populated parts of the world. In March 1993, the situation was even worse. The World Meteorological Organization reported that ozone levels over much of the Northern Hemisphere were 10 to 20 percent below normal—another temporary decline, but a very significant one.

Furthermore, in the last decade, the total amount of stratospheric ozone worldwide has decreased by more than 3 percent. This loss isn't a seasonal variation, but a steady, continuing decline. Even though stratospheric ozone is continually reformed by the ozone cycle, there is so much chlorine in the stratosphere today— 30 percent more than in 1985—that ozone molecules are being destroyed faster than they can be created.

And no matter what we do, chlorine levels will increase (and ozone levels will decrease) for many more years to come. The

Volcanoes like Mt. Pinatubo can shoot millions of tons of dust and acid particles into the stratosphere when they erupt. These particles can act like the ice crystals in PSCs and lead to ozone destruction in any part of the ozone layer, not just over polar regions.

reason for this is that there are millions of tons of CFCs in the environment today, just beginning their slow journey up to the stratosphere. Even if the production of CFCs ended tomorrow, chlorine levels in the stratosphere wouldn't begin to drop for many decades.

As the amount of chlorine in the stratosphere continues to increase—and as ozone depletion gets worse—more harmful UV radiation will be able to pass through the ozone layer and reach the earth's surface. Many scientists believe that the increased amount of UV radiation may seriously affect life on earth. Although no one knows exactly what will happen, some predictions have been made about the sorts of problems the world could be facing in the years ahead.

For people, a thinning ozone layer is a health risk. As the amount of harmful UV radiation reaching the surface of the earth increases, people's chances of developing skin cancer and other medical problems will increase as well. Skin cancer often takes many years to develop. Medical researchers estimate that because of ozone depletion, 12 million people in the United States will develop life-threatening forms of skin cancer over the next 50 years. If there were no ozone depletion, only 500,000 people would be expected to develop such skin cancers during that same period of time.

Our eyes are also sensitive to UV radiation. Repeated exposure to UV rays can cause cataracts to form in people's eyes. Cataracts lead to blindness if left untreated. As more and more harmful UV rays reach the earth through a thinning ozone layer, chances are that more people will suffer from cataracts in the future.

Increased UV radiation may also weaken our immune systems. The body relies on the immune system to fight infection. No one knows what a weakening of this important system could mean in terms of peoples' ability to fight off disease.

There is no doubt that damage to earth's ozone layer could create a serious risk to human health. But people, at least, can take steps to protect themselves from UV radiation. We can, for example, stay out of the sun, use sunscreen, and wear protective clothing. In some ways, we should be more concerned about what increased amounts of harmful UV radiation could mean for the animals, plants, and ecosystems around us.

Most animals have some kind of body armor that shields their

A tortoise's shell shields its body from UV radiation.

skin from UV radiation. Shells, scales, fur, and feathers all help block most harmful UV rays. But will that be enough? In southern Argentina, farmers are complaining that abnormally large numbers of sheep are developing eye problems. Are higher levels of UV radiation the cause? Because of ozone depletion, will earth's animals—from wild species to household pets—have problems fighting infection and disease? At this point, no one has answers to these troubling questions.

Plants will actually be threatened most directly by increased amounts of harmful UV rays. Because plants need light for photosynthesis, they must be exposed to sunlight and so can't avoid being exposed to UV rays. The types of plants growing in the world today have adapted to the levels of UV radiation that were striking the earth's surface before stratospheric ozone depletion began. Even though plants have some natural defenses against UV radiation, we don't know how much more UV radiation plants can tolerate.

If plants are seriously damaged because of ozone depletion, food supplies could be in jeopardy. In Australia, for example, agricultural scientists report that crops of wheat, sorghum, and peas are already showing the effects of higher levels of UV radiation. All in all, the impact of ozone destruction on the world's food supply could be a far more serious problem for most people than an increased risk of developing skin cancer or cataracts.

There is no doubt that ozone destruction is a complex environmental problem. But there is a long-term solution. The way to stop ozone destruction is to stop releasing ozone-destroying substances into the environment.

Many kinds of plants, including those we depend on for food, could be damaged as a result of ozone depletion.

A DuPont technician evaluates an alternative refrigerant in one of the company's test cars.

CHAPTER

7

Solving the Problem

Getting rid of CFCs and other ozone-destroying chemicals might seem to be a fairly simple thing to do. But for many millions of people in dozens of countries, CFCs are a part of everyday life. Take a look around your home. The refrigerator in your kitchen uses CFCs to keep food cold. Air conditioners, which many Americans have in their houses and cars, use CFCs for cooling, too. Sofa cushions, pillow stuffing, carpet padding, and foam insulation are all made using CFCs. In short, CFCs, or products made with them, are found all over the average American house or apartment.

Now think of all the houses or apartment buildings in your neighborhood. There are probably several dozen at least. How many in your town or city? A hundred? A hundred thousand? A million or more? Try to imagine all the houses and apartment buildings in all the cities in the entire country—that's many millions of buildings. And if you looked, it's very likely that you'd discover CFCs somewhere inside almost all of them.

Solving the problem of ozone destruction means getting rid of all those CFCs and CFC-made products that have become a normal part of modern-day life. It also means that everyone,

Products that were made with or contain CFCs include Styrofoam packing materials, cups, and egg cartons, as well as some types of aerosol sprays.

everywhere, needs to help make that happen, because ozone destruction is a global problem.

The information that scientists have gathered in just the past few years about ozone holes and ozone depletion proves that damage to our planet's ozone shield doesn't affect just one country or even one hemisphere. When ozone-destroying chemicals escape into the atmosphere—no matter where they originally come from—they damage the entire ozone layer, all around the globe. That means that all nations, big and small, must work together to rid the world of CFCs.

Countries began cooperating on the ozone depletion problem in the late 1980s. By then, many people had realized that unless something was done about CFCs, an ozone hole would continue to form over Antarctica each spring, and the ozone layer would continue to get thinner all around the world. On September 16, 1987, representatives from 24 nations, including the United States and other large industrialized countries, met in Montreal, Canada, to talk about a global CFC-reduction plan. After many discussions, an agreement was written that outlined a plan to gradually reduce the production of the five most common CFCs. This international treaty to protect the ozone layer was called the Montreal Protocol.

As scientists gathered more information about ozone destruction, however, it soon became obvious that gradually cutting back on a few types of CFCs wasn't going to be enough to solve the problem. So nations returned to the conference table. In 1990, representatives from nearly 90 countries met to revise the Montreal Protocol. They agreed to add 11 more CFCs and related chemicals to the Protocol's original list and pledged to ban all these substances by the end of the 20th century.

Less than a year later, Antarctica's ozone hole was larger than ever before. And in the spring of 1992, the possibility of ozone holes forming over North America, Europe, and Asia became a growing concern. With ozone disappearing at a much faster rate than anyone ever predicted, the Protocol's timetable for phasing out CFCs was sped up even more. Ninety-three countries agreed to phase out CFCs by the end of 1995.

Another set of ozone-destroying chemicals has been added to the Protocol, too. These include halons (used in fire extinguishers), methyl chloroform (used in dry cleaning clothes), carbon tetrachloride (used as a pesticide and added to paints), and methyl bromide (another pesticide ingredient). These chemicals attack stratospheric ozone in slightly different ways than CFCs do, and they aren't as plentiful in the atmosphere as CFCs are today. Nevertheless, they are a serious threat to the ozone layer and will probably be banned by 1995 or 1996.

Will the Montreal Protocol help save the ozone layer? The answer is a definite yes. If CFCs and other ozone-destroying chemicals are phased out according to the latest version of this international agreement, scientists expect to see a gradual improvement in the condition of the ozone layer beginning just after the turn of the century. Even then, it will take another 50 years or so for the amount of chlorine in the stratosphere to fall to the level it was before Antarctica's ozone hole first appeared.

Living without CFCs

In most industrialized countries, CFCs have been part of everyday life for a long time. Ridding the world of these chemicals in just a few years is a challenging problem that requires creative solutions.

One fairly simple step that has been taken to reduce the quantities of CFCs produced each year is to recycle these chemicals. In the United States, roughly 82 million cars are equipped with air conditioners. Not long ago, if a car's air conditioner needed to be repaired, the CFCs it contained were simply drained out and allowed to evaporate into the air. Today the procedure is quite different. Many service stations now use equipment designed to carefully extract CFCs from car air conditioners so they can be used over and over again.

Progress has been made with CFCs in refrigerators, too. At least 15 percent of the CFCs used worldwide are found in refrigerators. In the past, most worn-out refrigerators were hauled to

Old discarded refrigerators are a source of CFCs that escape into the environment and begin rising into the stratosphere.

landfills, where the CFCs they contained eventually leaked out and evaporated into the air. Germany was the first country to set up a national program for recovering and recycling CFCs from old refrigerators before they are discarded. Several other European countries are following Germany's example and are designing their own refrigerator-cleanup strategies.

Recycling on a smaller scale can also play a role in cutting down on CFC production. Foam plastics, such as disposable plates, hot drink cups, and packaging materials, were all once made with CFCs. Although it is true that some of these products are made

In many industrialized countries, CFCs used as coolants in air conditioners and refrigerators are now carefully collected and recycled.

without CFCs today, it's not always easy to tell for sure which ones are CFC-free. By reusing or doing without these materials, we can all help reduce the need to make more of them. Every little savings helps cut down on the production of new CFCs.

Recycling CFCs is certainly a step in the right direction, but the goal is to do away with CFCs entirely. For the past few years, companies that manufacture CFCs have been experimenting with new chemicals that they hope will take the place of CFCs and be harmless to the ozone layer.

Some of these substitutes are better than others. **Hydrochlorofluorocarbons (HCFCs),** for example, are very similar to

CFCs and can be substituted for them in many ways, especially in refrigerators. HCFCs tend to break down more rapidly in the atmosphere than CFCs do. The companies that manufacture HCFCs claim that because of this difference, most HCFC molecules should break down in the troposphere. But the manufacturers admit that some HCFC molecules will eventually reach the stratosphere. Once there, HCFCs will behave just like CFCs by breaking apart and releasing the ozone-destroying chlorine they contain. In short, HCFCs are a little less damaging than CFCs, but they still harm the ozone layer. As a result, HCFCs are not truly acceptable CFC-substitutes. They will probably be used for a while, but most countries have agreed to ban them by 2020.

Since it is chlorine in CFCs that attacks ozone in the ozone layer, chlorine-free compounds are potentially the best CFC substitutes. **Hydrofluorocarbons (HFCs)** are one group of chlorine-free compounds being tested as substitutes for several different types of CFCs. HFCs don't appear to be harmful to ozone. However, they do burn and they may be toxic, so there are doubts about how safe they will be.

Several companies have discovered natural substitutes for the CFCs they once used. AT&T now uses a chemical found in cantaloupes, peaches, and plums to replace a type of CFC the company once used for cleaning computer chips. This naturally occurring chemical—which smells like ripe bananas—cleans the chips as well as CFCs and is not only safe for the ozone layer, but it is safe for people to handle too. Another U.S. company has replaced the CFCs they used in the manufacturing of electronic circuit boards for military fighter jets with ordinary lemon juice.

Given time, substitutes for most CFCs will probably be found. The pressure to find safe substitutes quickly, however, has many scientists worried. New chemicals need to be tested, sometimes

One of the major uses for CFCs in years past was to clean electronic circuit boards and computer parts. Several substitutes for these CFCs have been found that will do the same job without harming the ozone layer.

for many years, to find out for sure if they are really safe. After all, when CFCs were first invented, nearly everyone thought they were harmless. Could one or more of the CFC-substitutes being introduced today turn out to be the cause of a new environmental problem 10, 20, or 50 years from now? No one really knows. But it is a concern that often is pushed aside in the rush to replace CFCs.

Another way to reduce production and use of CFCs is to

redesign equipment and processes so that CFCs are no longer needed. Take refrigerators as an example. Most refrigerators today operate using a piston pump and a coolant (a CFC). When the pump runs, it compresses the coolant, changing it from a gas into a liquid. The liquid moves out of the pump into a network of small tubes. As it moves through the tubes, the liquid changes back into gas again. In the process, it absorbs heat from its surroundings and in so doing cools the inside of the refrigerator.

Several research groups are experimenting with new types of refrigerators that don't use CFCs. One design uses a special pump that was invented in 1816 by Robert Stirling. The coolant inside a Stirling pump is helium. Helium is safe for the environment. It is the same lighter-than-air gas used to fill ozonesondes and other high-flying balloons—even the balloons you can buy at malls and amusement parks. A Stirling pump cools more efficiently than a CFC-run refrigeration system. It also runs silently.

Finally, we must remember that even when CFC-substitutes and CFC-free technologies become a reality, huge quantities of CFCs will still be stockpiled around the world. These CFCs will need to be destroyed so that there is no chance that they can ever get into the environment.

Several Japanese researchers recently designed a "CFC disintegrator" that works by heating CFCs with water to roughly 10,000°C (18,000°F). Not only are the CFCs destroyed in this process, but the by-products are harmless. Other types of CFC-destroying systems are also being tested in other countries.

Who Will Pay?

Chlorofluorocarbons have made it possible for people living in industrialized countries to enjoy the comforts and conveniences of modern life. Because industrialized countries have produced

Can We Fix It?

Eliminating CFCs from our lives and from the environment is going to take many years. Isn't there some way that the world could save all that time and effort and expense and just fix the ozone layer? Couldn't we just send some replacement ozone up into the stratosphere and solve the problem?

Sounds like a great idea at first. But how would you get the ozone up to the stratosphere? Several scientists have thought about what it would take to carry out such an operation. Using a 747 airplane that could carry 100 tons of ozone, they calculated that the plane would have to make 350,000 trips into the stratosphere to replace just 10 percent of the ozone there. Both the energy required for such a job and the cost would be enormous.

Furthermore, even if we could get the replacement ozone up into the stratosphere, it wouldn't solve the problem of ozone destruction: the CFCs would still be there. The chlorine in those CFCs would destroy the replacement ozone just as it is destroying the ozone that is there already.

What about trying to inactivate CFCs, especially over Antarctica, where damage to the ozone layer is currently

and used the most CFCs, they are responsible for most of the damage that's been done to the ozone layer so far.

Giving up CFCs and still maintaining a comfortable, modern life-style will be challenging and expensive. Wealthy, industrialized nations, like the United States, Japan, and Germany, can greatest? Could a fleet of jet aircraft fly into the Antarctic ozone hole and spray some sort of chemical into the stratosphere that would combine with CFCs and prevent them from attacking ozone? It's an interesting idea, but one that probably would be impossible to carry out. Planes loaded with the inactivating chemical would have to fly hundreds of missions into the hole at just the right time and release just the right amount of chemical in just the right place. Even then there would be no guarantee that ozone destruction would stop. In fact, if the chemicals didn't do exactly what they were supposed to do, there would be a good chance that they might make the situation even worse.

Thinking creatively about how to solve tough problems is a normal part of science. Unfortunately, sometimes newspaper and magazine articles present scientists' ideas for possible solutions to environmental problems in a way that makes these ideas sound like guaranteed solutions. Unless you read such articles carefully, you might assume that the environmental problems have been solved and that there is nothing more for people to worry about. Quick fixes may work in science fiction novels, but at least for now, the only real way to protect the ozone layer is to stop producing and using CFCs and other ozone-destroying chemicals.

probably afford to make this change. But what about developing countries such as India, China, and many African and Asian nations? The people in these countries are just beginning to enjoy some of the conveniences—like refrigerators—that CFC-technology can bring. China is a developing nation that has more than 800 million inhabitants. Up until just a few years ago, having a refrigerator was a luxury only the wealthiest people in China could afford. Today refrigerators are much more affordable, and the demand for them is tremendous. In the early 1980s, China produced 500,000 refrigerators annually. Now they make 8 million each year, and the coolants in all these new refrigerators are CFCs.

China and other developing countries say they can't afford to replace CFCs with more expensive ozone-safe substitutes. And they

China is beginning to produce automobiles, complete with CFC-containing air conditioners.

As more people in developing nations are able to afford the luxuries of modern life, such as automobiles, refrigerators, and air conditioners, the greater the need to find ozone-safe, affordable substitutes for CFCs as quickly as possible.

don't feel they should have to. Is it fair, they ask, for rich, industrialized countries responsible for today's ozone depletion to force poorer countries to share the costs of solving the CFC problem? But unless all nations agree to help rid the world of CFCs, any international plans to protect the ozone layer will fail.

In 1990, as part of the revisions made to the Montreal Protocol, a $240 million fund was set up by industrialized countries to help poorer nations pay for the cost of changing over to CFC-free technologies. This ozone fund has encouraged several developing countries to join in the effort to save the ozone layer. But there is still a long way to go before the entire world is united in working to protect the ozone layer—the irreplaceable atmospheric shield that protects life on earth.

Because of ozone depletion, the risk of damage from harmful UV radiation is higher than ever before.

PROTECT YOURSELF

The Montreal Protocol and other international agreements have helped reduce the amount of ozone-destroying chemicals being produced worldwide. And if people everywhere continue to work hard and take the problem seriously, the day will come when we finally stop releasing these chemicals into the environment altogether.

But it is important to remember that even when that happens, ozone depletion is going to continue for many years afterward. Tons of CFCs and other ozone-destroying chemicals are just beginning their slow journey up to the stratosphere, where they will attack the ozone layer. Ozone holes, a thinning ozone layer, and the dangerous UV radiation that can get through our damaged ozone shield are problems that are going to be with us for some time to come.

Being out in the sun means being exposed to UV radiation that can damage your skin, other parts of your body, and your health. Getting a lot of sun has never been a good idea, even when the ozone layer was intact. So in a way, ozone depletion simply adds to the risk that was already there. But that extra risk shouldn't be taken lightly. Medical researchers say that the chances of developing UV-related health problems will go up as the amount of ozone in the stratosphere goes down. The fact that the ozone layer is somewhat thinner than it was a few years ago means that your chances of getting skin cancer, for example, are somewhat greater than they used to be. And if an ozone hole like the one that forms each spring over Antarctica ever appears over where you live, the risk would be far, far greater.

The UV rays that strike the earth aren't like death rays that do their damage instantly. You won't be zapped one day and develop skin cancer the next. The damage caused by UV radiation is subtle and gradual, and it often takes years to show up. Medical studies have shown that the amount of UV radiation that you are exposed to when you are young affects how healthy your skin (and other parts of your body) will be when you are older. So it is important to play it safe with the sun and UV radiation.

So what do you do? Should a thinning ozone layer keep you inside all day, hiding from the sun? Not at this point. But you should take precautions against the increased amounts of harmful UV rays that are reaching the earth.

1) Avoid direct sunlight between 10:00 A.M. and 2:00 P.M. (11:00 A.M. and 3:00 P.M. during daylight savings time). This is the period of the day when the sun is highest overhead and when the UV radiation it emits strikes the earth most directly and intensely. And don't let a cloudy day fool you. Ultraviolet rays can easily pass through clouds, so even when it isn't sunny outside, you can still be exposed to a lot of UV radiation during the middle of the day.

2) Wear protective clothing when you are out in the sun. UV radiation can't pass through tightly woven fabrics. Long sleeves and pants will protect the skin on your arms and legs. Get into the habit of wearing a hat, especially the type with a wide brim that will protect the tops of your ears and the skin on your neck. Take a tip from people in Australia and New Zealand. The fact that Antarctica's ozone hole is so close to these two countries has made people living there much more aware of the problems UV radiation can cause. Today it's hard to find Australians or New Zealanders who don't wear hats when they are out in the sun!

Protect yourself from harmful UV rays with sunscreen and a hat, making sure to cover your ears and the back of your neck. Sunglasses to protect your eyes are also important.

3) Use sunscreen, preferably one that has a Sun Protection Factor (SPF) of at least 15, on the parts of your body that are exposed to the sun. Put on sunscreen at least 20 minutes before going out in the sun. Reapply it after going swimming or if you've been sweating a lot. Get into the habit of taking sunscreen along on outings so its protection will always be handy.

4) Ultraviolet radiation can damage your eyes and cause cataracts to form in them when you are older. Protect your eyes from UV rays by wearing sunglasses that are specially treated to absorb these harmful rays.

Keep in mind a simple saying that people in Australia use: "Slip, Slop, Slap!"—slip on a shirt, slop on some sunscreen, and slap on a hat (and sunglasses). If you follow these simple steps, you'll be doing a lot to protect yourself from harmful UV radiation. It's worth the effort.

Glossary

Antarctic polar vortex—an enormous ring of moving air created by powerful stratospheric winds that blow in a clockwise direction around the Antarctic continent during the winter months

chlorine monoxide (ClO)—a highly reactive compound that is involved in ozone-destroying reactions in the stratosphere. The presence of ClO is a telltale sign that stratospheric ozone destruction is occurring.

chlorofluorocarbons (CFCs)—synthetic chemicals made up of different combinations of chlorine, fluorine, and carbon. Used for many industrial purposes, CFCs destroy stratospheric ozone that protects the earth from harmful ultraviolet radiation coming from the sun.

diatoms—a type of phytoplankton common in most of the world's oceans. Diatoms are found in a variety of shapes and sizes, but most are too small to be seen with the naked eye.

electromagnetic radiation—energy that travels through space in the form of waves. The energy given off by the sun is electromagnetic radiation.

electromagnetic spectrum—the range of wavelengths of electromagnetic radiation, from the shortest-wave gamma rays to the longest radio waves

hydrochlorofluorocarbons (HCFCs)—synthetic chemicals proposed as short-term substitutes for chlorofluorocarbons (CFCs). HCFCs are very similar in composition to CFCs.

hydrofluorocarbons (HFCs)—synthetic, chlorine-free chemicals proposed as substitutes for some types of CFCs

lidar—short for "light radar," a scientific instrument that emits pulses of laser light and is used to measure molecules and particles in the atmosphere

mesosphere—the third region of the earth's atmosphere, extending to about 80 kilometers (50 miles) above the surface

ozone (O_3)—a strong-smelling, pale blue gas that is formed naturally in the stratosphere by the action of ultraviolet radiation on oxygen molecules. A molecule of ozone is made up of three atoms of oxygen.

ozone layer—a layer of ozone gas in the stratosphere that shields the earth from most of the harmful ultraviolet radiation coming from the sun

ozonesondes—balloon-carried devices used to measure ozone in the atmosphere

photosynthesis—the process by which green plants use energy from sunlight to produce their own food. As part of this process, plants take in carbon dioxide and water, and release oxygen.

phytoplankton—minute, usually one-celled planktonic plants. Countless numbers of phytoplankton drift through the world's oceans and form the base of marine food webs.

polar stratospheric clouds (PSCs)—large, diffuse, ice-particle clouds that form in the stratosphere over polar regions

reservoir molecules—molecules in the atmosphere that bind with atoms or other molecules and prevent them from participating in chemical reactions

stratosphere—the second region of the earth's atmosphere, extending to roughly 48 kilometers (30 miles) above the surface. The ozone layer is contained within this region of the atmosphere.

thermosphere—the fourth and outermost region of the earth's atmosphere, extending out about 965 kilometers (600 miles) above the surface

troposphere—the innermost region of the earth's atmosphere, extending out about 11 kilometers (7 miles) above the surface

ultraviolet (UV) radiation—energy waves with wavelengths ranging from about 0.005 to 0.4 micrometers on the electromagnetic spectrum. Most ultraviolet rays coming from the sun have wavelengths between 0.2 and 0.4 micrometers. Much of this high-energy radiation is absorbed by the ozone layer in the stratosphere.

wavelength—the distance from one high point of a wave to the next high point (or from one low point to the next low point). Waves of electromagnetic radiation are often measured in terms of their wavelengths.

zooplankton—small, planktonic animals that drift through the world's oceans. Many types of zooplankton eat phytoplankton.

The ice runway on the sea ice near McMurdo Station, Antarctica

Index

manufacturing foam plastics, CFCs in, 37, 89, 90, 93-94
mesosphere, 16, 17
methyl bromide, 92
methyl chloroform, 91
Molina, Mario J., 38-40, 43
Montreal Protocol, 91-92, 101, 103

National Aeronautics and Space Administration (NASA), 30, 32, 80
National Oceanic and Atmospheric Administration (NOAA), 9-12, 32
National Ozone Expedition (NOZE), 32, 33
Neale, Patrick, 69-72
Nimbus 7 satellite, 30
nitrogen, 15, 43

oxygen (O_2), 15, 17-18, 22-23, 40-41, 42
ozone (O_3), 12, 15, 18, 22-23, 40-41, 42-43, 51, 54, 59, 98; as form of pollution, 24-25; "replacement," 98
ozone cycle, 18, 22-23, 26, 42-43
ozone depletion, 12-13, 28, 29-32, 39, 40-41, 46-48, 54, 61, 75, 77-78, 81, 83, 89-90
ozone layer, 12, 16, 17, 18, 26-27, 41, 43, 48, 51, 61, 63, 90, 98, 103; "hole" in, 12-13, 27, 32, 33, 34, 39, 43, 53, 61, 65, 72, 77-78, 91, 103; thinning of, 13, 43, 75, 103
ozonesondes, 55-60, 79

Palmer Station, Antarctica, 9, 66-68
penguins, 62, 64, 65, 74, 77
photosynthesis, 64, 65, 72, 86
phytoplankton, 63, 64-72, 74, 77
plants, effect of UV rays on, 86-87
plastics, CFCs in. *See* manufacturing foam plastics
polar stratospheric clouds (PSCs), 49-52, 53, 78, 79, 80
polar vortex, 48, 49-52, 53, 77-78, 79, 80, 81

refrigerators, CFCs in, 37, 88, 92-93, 94, 97, 99-100
reservoir molecules, 43, 51-52, 53, 78, 80
Ross Island, 8, 9, 68, 69
Rowland, F. Sherwood, 38-40, 43

Sanders, Ryan, 33
satellites, use in studying ozone layer, 30, 31, 32, 60, 80, 82
seabirds, 65, 74
seals, 64, 65, 75
skin cancer, role of UV rays in, 85, 103-104
Smith, Joe, 7-12
Solomon, Susan, 32
Stirling pump, 97
stratosphere, 16, 17, 18, 29, 32, 39, 41, 48, 49, 50
Styrofoam, CFCs in. *See* manufacturing foam plastics

thermosphere, 16, 17

Photo Acknowledgments

The photographs and illustrations in this book are reproduced through the courtesy of: National Science Foundation, pp. 2, 8, 62, 65 (inset), 75; Stuart Klipper, pp. 6, 10; Brian Liedahl, pp. 9, 16, 23, 27, 41, 42, 49, 53; © Rebecca L. Johnson, pp. 11, 28, 44, 55, 56, 69, 71, 109; National Oceanic and Atmospheric Association, p. 13; NASA, pp. 14, 31, 34, 35, 50, 61, 76, 82; Laura Westlund, pp. 19, 20–21, 65; © Barbara Laatsch-Hupp/Laatsch-Hupp Photo, p. 25; Noze Science Team, p. 33; © Richard B. Levine, pp. 36, 90; University of California Irvine, p. 38; Ryan Sanders, pp. 47, 59; University of Wyoming, pp. 57, 58, 79; Deneb Karentz, pp. 64, 67, 70; Visuals Unlimited/ © Charles Preitner, p. 66; Visuals Unlimited/ © Frank T. Awbrey, p. 73; David Chittenden, p. 74; Official USAF photo by SRA Paul Davis, p. 84; R.E. Barber, p. 86; USDA, p. 87; DuPont, pp. 88, 94, 96; Visuals Unlimited/ © Bob Newman, p. 93; OPIC, p. 100; Steve Feinstein, p. 101; Oregon Tourism Division, p. 102; American Cancer Society, p. 105.

Front cover photograph courtesy of Dave Hopmann. Back cover photograph courtesy of National Science Foundation.